THEORETICAL CHEMISTRY
Advances and Perspectives

Volume 3

Contributors to This Volume

Henry Eyring
Mu Shik Jhon
R. D. Murphy
Jack Simons
Frank H. Stillinger
Frank Weinhold

THEORETICAL CHEMISTRY: Advances and Perspectives

VOLUME 3

EDITED BY

HENRY EYRING

Department of Chemistry
University of Utah
Salt Lake City, Utah

DOUGLAS HENDERSON

IBM Research Laboratory
Monterey and Cottle Roads
San Jose, California

ACADEMIC PRESS New York San Francisco London 1978
A Subsidiary of Harcourt Brace Jovanovich, Publishers

ACADEMIC PRESS, INC.
111 Fifth Avenue, New York, New York 10003

United Kingdom Edition published by
ACADEMIC PRESS, INC. (LONDON) LTD.
24/28 Oval Road, London NW1

LIBRARY OF CONGRESS CATALOG CARD NUMBER: 75–21963

ISBN 0–12–681903–3

PRINTED IN THE UNITED STATES OF AMERICA

Contents

A Model of the Liquid State. Three Phase Partition Functions

Mu Shik Jhon and Henry Eyring

Structure of Fluid ^4He

R. D. Murphy

Proton Transfer Reactions and Kinetics in Water

Frank H. Stillinger

List of Contributors

Numbers in parentheses indicate the pages on which the authors' contributions begin.

Henry Eyring (55), *Department of Chemistry, University of Utah, Salt Lake City, Utah*

Mu Shik Jhon* (55), *Institut de Biologie Physico Chimique, University de Paris, Paris, France*

R. D. Murphy (143), *Department of Physics, University of Missouri-Kansas City, Kansas City, Missouri*

Jack Simons (1), *Department of Chemistry, University of Utah, Salt Lake City, Utah*

Frank H. Stillinger (177), *Bell Laboratories, Murray Hill, New Jersey*

Frank Weinhold (15), *Theoretical Chemistry Institute and Department of Chemistry, University of Wisconsin-Madison, Madison, Wisconsin*

* Present address: Korea Advanced Institute of Science, P.O. Box 150, Cheong Ryang Ri, Seoul, Korea.

Preface

In a sense, theoretical chemistry has existed as long as scientists have tried to understand chemical phenomena. However, it has been only recently that theoretical chemistry has grown into a mature field. Three developments, all relatively recent, have spurred this growth. First, at the end of the nineteenth century the foundations of statistical mechanics were laid. As a result, the bulk properties of chemical systems could be calculated from their microscopic dynamics. Second, during the first third of the twentieth century, quantum mechanics was developed, giving a satisfactory theory for the microscopic dynamics of chemical systems. Finally, in the past two decades, fast electronic computers have made accessible the full richness of quantum and statistical mechanics for the theoretical description of complex chemical systems.

Despite the maturity of theoretical chemistry, there are very few journals or review series devoted to all aspects of this field. It is hoped that this serial publication will fill, in part at least, this gap. Articles concerning all aspects of theoretical chemistry will be published in these volumes. Articles concerning experimental chemistry which pose or answer questions of theoretical interest may also be published from time to time.

The quantum mechanical study of stable molecular anions has been retarded by the lack of highly accurate methods for treating electron correlation effects in these species. In this volume, Simons reviews recent developments which have made the *ab initio* investigation of molecular ions feasible. Weinhold discusses classical equilibrium thermodynamics. He surveys recent work in this field, sometimes incorrectly considered a closed subject, which reveals an underlying geometrical structure of unexpected beauty and simplicity and discusses the classical thermodynamic laws in light of these developments, emphasizing how each law contributes to the way the formalism ultimately works.

Jhon and Eyring update their work on the significant structure theory of liquids. Models of the liquid state, to be most useful, must agree with and illuminate machine calculations for both the relatively simple molecules and the restricted numbers of molecules now accessible by such means and should also provide the framework for understanding and treating more complicated systems. Significant Structure Theory has as its ambitious program the treatment of solid liquid and gas thermodynamic and transport properties using a single partition function. The present chapter provides an assessment of the progress made toward this difficult long range goal.

Murphy discusses the structure of fluid ^{4}He by means of two approaches, one valid at zero temperature and the other valid at nonzero temperatures. The radial distribution function is calculated for both approaches by Monte Carlo methods, and the connection between the two approaches is discussed. Finally, Stillinger surveys the various types of proton transfer reactions in water. Such processes play a fundamental role in life processes at the molecular level. Stillinger shows how the great speed of such reactions is intimately associated with the unusual hydrogen-bond structure that characterizes liquid water. Specific attention is focused on the dissociation of water molecules themselves (the most elementary proton transfer reaction), on solvation structure of the H^+ and OH^- ions thus formed, and on the electrical conduction mechanisms for these ions. In order to facilitate the overall analysis, a general formalism is developed in the chapter to describe rates of chemical reaction by means of autocorrelation functions for microscopic fluxes.

HENRY EYRING
DOUGLAS HENDERSON

Contents of Previous Volumes

Volume 1

Volume 2

Theoretical Studies of Negative Molecular Ions

Jack Simons†
Department of Chemistry
University of Utah, Salt Lake City, Utah

I. Introduction

In a recent review article on negative molecular ions (Simons, 1977), coverage was concentrated on *results* of quantum chemical calculations on specific anions; very little discussion was devoted to the methods used to carry out these calculations. Because the present series is devoted to covering developments in theoretical chemistry, it is perhaps more appropriate to direct the present review toward explaining the quantum mechanical *techniques* that are currently used to study molecular anions. A more complete overview of the current status of theoretical research on negative ions can be obtained by reading this article and the author's other review (Simons, 1977).

† Camille and Henry Dreyfus Teacher Scholar.

To understand why one needs rather sophisticated methods to study negative ions reliably, one need only to consider the results of our calculations (Smith *et al.*, 1974) on the vertical ($R = 1.718$ a.u.) electron detachment energy ($X^1\Sigma^+ OH^- \rightarrow X\,^2\pi_i\,OH$) of OH^-. Using an atomic orbital basis consisting of twenty Slater-type orbitals (STO's), we obtained a Koopmans' theorem approximation to the electron detachment energy equal to 3.06 eV. The energy difference between two separate self-consistent field (SCF) calculations (ΔSCF), one on OH^- and one on OH, carried out within the same basis was equal to −0.20 eV; the difference between Koopmans' theorem and ΔSCF represents the effects of allowing the orbitals to relax upon removal of the π electron. Finally, our best computed energy difference, which contains effects of electron correlation through third order, was 1.76 eV, which is in good agreement with both Branscomb's early experiments (Branscomb, 1966) and Lineberger's more recent laser detachment results (Hotop *et al.*, 1974). The difference between 1.76 eV and the ΔSCF value of −0.2 eV represents the effects of electron correlation. These effects are indeed as large as the entire electron affinity of OH; moreover, this result is not atypical.

Because the treatment of both orbital relaxation and electron correlation effects in a sufficiently rigorous manner is an absolute necessity in any reliable scheme for computing properties of anions, theoretical progress toward understanding negative ions has been made rather slowly. Quite simply put, it is difficult to include correlation effects to a high enough order to guarantee precision of ±0.2 eV in computed ion-neutral energy differences. In Berry's 1968 review article (Berry, 1969), his assessment of the state of quantum chemical research on anions involved briefly mentioning the works of Pekeris (1958) on H^-; Weiss (1968) on Li^-, Na^-, and K^-; Clementi (1964), Clementi and McLean (1964), and Clementi *et al.* (1964) on several atomic ions; Sinanoğlu and Oksuz (1968) on C^-, O^-, and F^-; Taylor and Harris (1963) on H_2^-; Wahl and Gilbert (1965) on halogen diatomics; and Cade (1967a,b) on OH^-, CH^-, SiH^-, SH^-, and PH^-. Therefore, in 1968 it would have been fair to say that negative molecular ions were not yet within the class of species that could conveniently be studied by existing quantum chemical methods. On the other hand, the development of modern laser technology was making available to the experimentalist new tools to use in carrying out high precision photodetachment and photoelectron spectroscopy studies of gas-phase anions. Thus, even in 1968, a great deal of experimental progress was beginning to be made. These experimental developments made a parallel development of theoretical methods and models aimed at better understanding negative ions a necessary and quite natural step in the scientific progress in this area.

In 1973 Simons and Smith published an article in which they attempted to use equations of motion (EOM) techniques to express the vertical electron affinity (EA) or detachment energy (DE) of a closed-shell species in a manner that treated orbital relaxation and electron correlation through third order in perturbation theory (the difference between the Coulombic interaction and the Hartree–Fock interaction being the perturbation). This developmental paper was followed by other formal papers by Simons (1974), Jørgensen (Jørgensen and Simons, 1975), and Jordan (Jordan *et al.*, 1976b), in which small deficiencies in the original theory were corrected and connection was made with the recent Green's function developments of Cederbaum (1973), Pickup and Goscinski (1973), Purvis and Öhrn (1974), and Freed (Tsui and Freed, 1975). The result of these papers was a method that permits the direct calculation of EA's and DE's of closed-shell species that are accurate through third order. Because the work of Doll and Reinhardt (1972) indicated that a second-order treatment of electron correlation effects was inadequate, this extension to third order was quite important. With this historical perspective in mind let us now turn our attention toward developing a unifying framework in whose terms we can understand the different techniques which are currently in use in this area of research.

II. The One-Electron Green's Function

The electron propagator or the one-electron Green's function has been used for some time (Cederbaum, 1973; Cederbaum *et al.*, 1971, 1973; Pickup and Goscinski, 1973; Purvis and Öhrn, 1974, 1975; Tsui and Freed, 1975; Griffing and Simons, 1975, 1976; Kenney and Simons, 1975; Andersen and Simons, 1976, 1977a,b; Griffing *et al.*, 1975; Jordan *et al.*, 1976a; Linderberg and Öhrn, 1967, 1973; Doll and Reinhardt, 1972; Schneider *et al.*, 1970; Yarlagadda *et al.*, 1973) in the study of electron spectroscopy. The advantages of using the electron propagator arise because the transition energies and the transition strengths are obtained directly as poles and residues of the propagator, respectively. Several alternative procedures for decoupling the equation of motion for the electron propagator have been developed. In this review we use the superoperator formalism of Goscinski and Lukman (1970) as the framework for our development.

The time-dependent Green's function $G_{ij}(t)$ is defined in terms of the spin-orbital annihilation (a_i) and creation (a_i^+) operators as follows:

$$G_{ij}(t) = (1/i)\{\langle a_i^+(t)a_j\rangle\theta(t) - \langle a_j a_i^+(t)\rangle\theta(-t)\}, \tag{1}$$

where $\theta(t)$ is the Heaviside step function and the bracket $\langle\ \rangle$ indicates an average over some reference state. By introducing the identity $\exp(iHt)a^+\exp(-iHt) = \exp(-i\hat{H}t)a^+$, where $\hat{H}$ is the superoperator Hamiltonian ($\hat{H}a^+ \equiv a^+, H$), and Fourier transforming Eq. (1), one obtains

$$\mathsf{G}(E) = (\mathbf{a}(E\hat{I} - \hat{H})^{-1}\mathbf{a}) \tag{2}$$

where $\hat{I}$ and $\hat{H}$ are the superoperator identity and Hamiltonian, respectively, and the **a** are the set of annihilation operators $\mathbf{a} = a_i$, which are arranged in a super row vector. The superoperator scalar product is defined in the conventional fashion, $(A \mid \hat{H} \mid B) \equiv \langle A^+(\hat{H}B) + (\hat{H}B)A^+ \rangle$. The superoperator resolvent $(E\hat{I} - \hat{H})^{-1}$ can be approximated via the inner projection technique in which case the propagator then takes the form

$$\mathsf{G}(E) = (\mathbf{a} \mid \mathbf{h})(\mathbf{h} \mid \hat{E}I - \hat{H} \mid \mathbf{h})^{-1}(\mathbf{h} \mid \mathbf{a}) \tag{3}$$

where **h** is a projection manifold which, if chosen to be complete and orthonormal, makes Eqs. (2) and (3) identical. The operator space

$$\{\mathbf{h}_1, \mathbf{h}_3, \mathbf{h}_5 \cdots\} = \{a_i, a_i^+ a_k a_1, a_i^+ a_j^+ a_k a_1 a_r \cdots\}, \qquad k > 1, i > j, k > 1 > r \cdots \tag{4}$$

spans the manifold **h**. We now discuss appropriate selections of **h** which, in conjunction with a choice of the reference state, ensure that the electron propagator is calculated correctly through third order in the electronic interaction. The development is limited to third order primarily because currently used methods have not been extended, except for selective summation of higher order terms, to fourth and higher order. Moreover, results (Griffing and Simons, 1975, 1976; Kenney and Simons, 1975; Andersen and Simons, 1976; Griffing *et al.*, 1975; Jordan *et al.*, 1976a; Andersen and Simons, 1977a,b) of third-order calculations seem to be very reliable (± 0.2 eV).

It is well known that the projection manifold $\mathbf{h}_1$, $\mathbf{h}_3$, in connection with the Hartree–Fock (HF) ground state, is able to give the electron propagator correct through second order in the electronic interaction. Such second-order calculations were first performed, for *ab initio* Hamiltonians, by Doll and Reinhardt (1972). Our experience tells us that second-order calculations of EA's are not sufficiently precise to be useful (except in well-understood special cases which are discussed in our earlier review). Thus, we choose to demonstrate how, using a correlated ground state and the same projection manifold, we are able to get the electron propagator correct through third order in the electronic interaction. The interested reader can then generate correct second-order results by simply ignoring the terms arising from correlation effects in the ground state.

The effect of including $\mathbf{h}_5$ in the projection manifold, where the HF ground state is used as reference state, has been discussed by Tyner-Redmon *et al.* (1975) and from their analysis it is clear that $\mathbf{h}_5$ introduces terms which are at least fourth order in the electronic interaction, independent of the choice of reference state. This conclusion is not easily reached in a rigorous derivation because the superoperator Hamiltonian $\hat{H}$ is not hermitian $[(\mathbf{h}_5 | \hat{H} | \mathbf{h}_1) = 0$, but $(\mathbf{h}_1 | \hat{H} | \mathbf{h}_5) \neq 0]$. However, Purvis *et al.* (1976) have shown that, although $\hat{H}$ is not rigorously hermitian, the nonhermitian terms are introduced by the improved (higher order) reference state. Thus, if one is careful to avoid the nonhermitian parts of $\hat{H}$, one can show, as Tyner-Redmon has, that the $\mathbf{h}_5$ operators cannot contribute to G^{-1} through third order. We therefore concentrate on using $\mathbf{h}_1, \mathbf{h}_3$ as our projection manifold in our search for a theory that is consistent through third order.

As the reference state in our analysis we use a correlated wave function given by

$$|0\rangle = N^{-1/2}\left\{1 + \sum_{p\delta} (K_\delta^p a_p^+ a_\delta) + \sum_{\substack{m>n \\ \alpha>\beta}} (K_{\alpha\beta}^{mn} a_m^+ a_n^+ a_\beta a_\alpha) + \sum_{\substack{m>n>p \\ \alpha>\beta>\delta}} (K_{\alpha\beta\delta}^{mnp} a_m^+ a_n^+ a_p^+ a_\alpha a_\beta a_\delta) + \cdots\right\}|\mathrm{HF}\rangle \tag{5}$$

where the $\mathbf{a}^+$ are a set of HF creation operators and where indices m, n, p, q $(\alpha, \beta, \delta, \gamma)$ refer to unoccupied (occupied) spin orbitals in the HF ground state and i, j, k, l, r are unspecified spin orbitals. We then take the correlation coefficients from Rayleigh–Schrödinger perturbation theory

$$K^p = \sum_{\substack{m>n \\ \alpha>\beta}} \{\langle p\alpha||mn\rangle\delta_{\delta\beta}\langle p\beta||mn\rangle\delta_{\alpha\delta} + \langle\beta\alpha||\delta m\rangle\delta_{np} - \langle\beta\alpha||\delta n\rangle\delta_{pm}\} \times \frac{\langle mn||\alpha\beta\rangle}{(\varepsilon_\delta - \varepsilon_p)(\varepsilon_\alpha + \varepsilon_\beta - \varepsilon_n - \varepsilon_m)} + \text{higher order terms in the electronic interaction} = K_\delta^p(2, 3, \ldots) \tag{6}$$

$$K_{\alpha\beta}^{mn} = \frac{\langle mn||\alpha\beta\rangle}{\varepsilon_\alpha + \varepsilon_\beta - \varepsilon_m - \varepsilon_n} + \text{higher order terms in the electronic interaction} = K_{\alpha\beta}^{mn}(1, 2, \ldots) \tag{7}$$

$$K_{\alpha\beta\delta}^{mnp} = K_{\alpha\beta\delta}^{mnp}(2, 3, \ldots) \tag{8}$$

where the numbers in the bracket indicate the orders in the electronic interaction contributing to K_δ^p, $K_{\partial\beta}^{mn}$, and $K_{\alpha\beta\delta}^{mnp}$. The ε_i indicate HF orbital energies and the two-electron integral $\langle mn|\alpha\beta\rangle$ refers to the charge densities $m\alpha$ and $n\beta$ and

$$\langle mn||\alpha\beta\rangle = \langle mn|\alpha\beta\rangle - \langle mn|\beta\alpha\rangle. \tag{9}$$

In the remainder of our analysis we employ the projection manifold $\{\mathbf{h}_1, \mathbf{h}_3\}$

where the $\mathbf{h}_3$ space for convenience has been redefined as

$$\mathbf{h}_3 = \{a_i^+ a_k a_l + \langle a_i^+ a_l \rangle a_k - \langle a_i^+ a_k \rangle a_l\}. \tag{10}$$

This choice of the subspace $\mathbf{h}_3$ ensures the space is orthogonal to $\mathbf{h}_1$

$$(\mathbf{h}_1 | \mathbf{h}_3) = 0, \tag{11}$$

even for a correlated reference state. We also have the following orthogonality relations:

$$(\mathbf{h}_1 | \mathbf{h}_1) = 1, \qquad (\mathbf{h}_3 | \mathbf{h}_3) = \mathsf{S}(0, 2, 3, \ldots). \tag{12}$$

Using Eqs. (11) and (12), Eq. (3) can be partitioned into the form

$$\begin{aligned} \mathsf{G}^{-1}(E) &= (\mathbf{h}_1 | E\hat{I} - \hat{H} | \mathbf{h}_1) - (\mathbf{h}_1 | \hat{H} | \mathbf{h}_3)(\mathbf{h}_3 | \hat{E}I - H | \mathbf{h}_3)^{-1}(\mathbf{h}_3 | \hat{H} | \mathbf{h}_1) \\ &= E\mathsf{1} - \mathsf{A} - \mathsf{B}\mathsf{D}^{-1}\mathsf{C} \end{aligned} \tag{13}$$

where the matrices ABC and D are defined as

$$\begin{aligned} \mathsf{A} &= (\mathbf{h}_1 | \hat{H} | \mathbf{h}_1) \qquad & \mathsf{B} &= (\mathbf{h}_1 | \hat{H} | \mathbf{h}_3) \\ \mathsf{D} &= (\mathbf{h}_3 | \hat{E}I - \hat{H} | \mathbf{h}_3) \qquad & \mathsf{C} &= (\mathbf{h}_3 | \hat{H} | \mathbf{h}_1). \end{aligned} \tag{14}$$

We are now prepared to make an order-by-order analysis of Eq. (13) in which we retain only those terms which are zeroth, first, second, or third order in the electronic interaction. Since, as will be seen shortly, the B and C matrices are at least of first order, we need to consider only that part of the D matrix that is zeroth and first order. This constrains the indices in the projection manifold $\mathbf{h}_3$ to be of the form $a_m^+ a_\alpha a_\beta \, \alpha > \beta$ or $a_\alpha^+ a_m a_n \, m > n$ since operators such as $a_m^+ a_n a_\alpha$ lead to matrix elements in the D matrix that are at least of second order. The resulting subspace $\mathbf{h}_3$ is thus identical to that used in calculating the electron propagator through second order in the electron repulsions (Doll and Reinhardt, 1972), although we now need to obtain the B and C matrices through second order, the D matrix through first order, and the A matrix through third order.

Previous attempts (Schneider *et al.*, 1970; Yarlagadda *et al.*, 1973) to obtain the electron propagator correct through third order have used the reference state

$$|0\rangle = N_0^{-1/2}\left(1 + \sum_{\substack{m>n \\ \alpha>\beta}} K_{\alpha\beta}^{mn} a_m^+ a_n^+ a_\alpha a_\beta\right) |\mathrm{HF}\rangle \tag{15}$$

where the $K_{\alpha\beta}^{mn}$ are determined from first-order Rayleigh–Schrödinger perturbation theory. In calculating the B, C, and D matrix elements correct through second and first order, respectively, no changes are obtained from considering the higher correlated ground state in Eq. (5). The matrix elements of **B**, **C**, and **D** are given, either for the reference state of Eq. (5) or for the state shown in Eq. (15) as

$$B_{i,\alpha m\beta} = -\langle im|\alpha\beta\rangle - \frac{1}{2}\sum_{p,q}\langle im|pq\rangle K_{\alpha\beta}^{pq} + \sum_{\gamma,p}[\langle i\gamma|p\alpha\rangle K_{\beta\gamma}^{mp} - \langle i\gamma|p\beta\rangle K_{\alpha\gamma}^{mp}] \quad (16)$$

$$B_{i,n\alpha m} = \langle i\alpha|mn\rangle + \frac{1}{2}\sum_{\gamma\delta}\langle i\alpha|\delta\gamma\rangle K_{\delta\gamma}^{mn} + \sum_{\gamma p}[\langle ip|\gamma n\rangle K_{\alpha\gamma}^{mp} - \langle ip|\gamma m\rangle K_{\alpha\gamma}^{np}] \quad (17)$$

$\mathbf{C}^{+} = \mathbf{B}$ (through second order).

$$D_{n\alpha m,\delta p\gamma} = 0, \quad (18)$$

$$D_{n\alpha m,q\beta p} = \delta_{nq}\delta_{\alpha\beta}\delta_{mp}(\varepsilon_m + \varepsilon_n - \varepsilon_\alpha - E) - \delta_{qn}\langle m\beta||p\alpha\rangle - \delta_{pm}\langle n\beta||q\alpha\rangle + \delta_{qm}\langle n\beta||p\alpha\rangle + \delta_{\alpha\beta}\langle mn||pq\rangle + \delta_{pn}\langle m\beta||q\alpha\rangle, \quad (19)$$

$$D_{\delta p\gamma,\alpha q\beta} = -\{\delta_{\delta\alpha}\delta_{pq}\delta_{\gamma\beta}(\varepsilon_p - \varepsilon_\delta\varepsilon_\gamma + E) - \delta_{\gamma\beta}\langle\delta q||\alpha p\rangle - \delta_{\delta\alpha}\langle\gamma q||\beta p\rangle + \delta_{\gamma\alpha}\langle\delta q||\beta p\rangle - \delta_{pq}\langle\delta\gamma||\beta\alpha\rangle + \delta_{\delta\beta}\langle\gamma q||\alpha p\rangle. \quad (20)$$

In the **A** matrix we need to include all terms up to third order. The **A** matrix elements obtained by using Eq. (15) as a reference state need to be modified by third-order terms that result from the interaction between the singly excited states of Eq. (5) and the HF ground state. The triply excited states of Eq. (5), which also result from a second-order Rayleigh–Schrödinger perturbation calculation, do not introduce third-order terms in **A**. We thus have to add to the **A** matrix elements given by Simons (Simons and Smith, 1973), which employed the reference state of Eq. (15), the terms δA_{ij}

$$\begin{aligned}\delta A_{ij} = &\sum_{\substack{\delta\beta\\pmn}} \frac{\langle jp||i\delta\rangle\langle\delta\beta||mn\rangle\langle mn||p\beta\rangle}{(\varepsilon_\delta - \varepsilon_p)(\varepsilon_\delta + \varepsilon_\beta - \varepsilon_m - \varepsilon_n)} \\ &+ \sum_{\substack{\delta\beta\\pmn}} \frac{\langle j\delta||ip\rangle\langle p\beta||mn\rangle\langle mn||\delta\beta\rangle}{(\varepsilon_\delta - \varepsilon_p)(\varepsilon_\delta + \varepsilon_\beta - \varepsilon_m - \varepsilon_n)} \\ &+ \sum_{\substack{\delta\alpha\beta\\pn}} \frac{\langle jp||i\delta\rangle\langle\delta n||\beta\alpha\rangle\langle\alpha\beta||pn\rangle}{(\varepsilon_\delta - \varepsilon_p)(\varepsilon_\alpha + \varepsilon_\beta - \varepsilon_p - \varepsilon_n)} \\ &+ \sum_{\substack{\delta\alpha\beta\\pn}} \frac{\langle j\delta||ip\rangle\langle\beta\alpha||\delta n\rangle\langle pn||\alpha\beta\rangle}{(\varepsilon_\delta - \varepsilon_p)(\varepsilon_\alpha + \varepsilon_\beta - \varepsilon_p - \varepsilon_n)}.\end{aligned} \quad (21)$$

The A matrix originally given by Simons (Simons and Smith, 1973) is

$$A_{ij} = \delta_{ij}\varepsilon_i + \sum_{k,l} \langle ik \mid jl \rangle F_{kl}, \tag{22}$$

where

$$F_{kl} = \sum_{\alpha > \beta, p} K_{\alpha\beta}^{pk} K_{\alpha\beta}^{pl} - \sum_{p<q, \alpha} K_{\alpha l}^{pq} K_{\alpha k}^{pq}. \tag{22a}$$

The results shown in Eqs. (16–22a) constitute a superoperator derivation of the electron propagator through third order. With these general results in hand, let us now turn to the analysis of various approximations that have been introduced for $\mathsf{G}(E)$.

III. Approximations to $\mathsf{G}(E)$

A. Cederbaum's Diagrammatic Method

In a series of papers aimed at developing an accurate and tractable quantum mechanical approach to principal and shake-up ionization energies and photoelectron intensities, Cederbaum (1973) and Cederbaum *et al.* (1971, 1973) has made use of Feynman diagram techniques to derive an approximation to the self-energy $\Sigma(E)$ appearing in the Dyson equation

$$\mathsf{G}^{-1}(E) = \mathsf{G}_0^{-1}(E) - \Sigma(E). \tag{23}$$

A comparison of Eqs. (13) and (23) shows that $\Sigma(E)$ can be written as $\mathsf{BD}^{-1}\mathsf{C} + \mathsf{A}^{(3)}$, where $\mathsf{A}^{(3)}$ is the third-order part of A if $\mathsf{G}_0^{-1}(E)$ is identified as $(E - \varepsilon_i \delta_{ij})$. In Figs. 1 and 2 are shown the diagrams that constitute the second- and third-order components of $\Sigma(E)$, respectively. The reader should consult the papers of Cederbaum for a review of the rules for writing, in terms of two-electron integrals and orbital energies, the values of Feynman diagrams. The (second-order) diagrams of Fig. 1 arise in our Eq. (13) for $G^{-1}(E)$ when the first-order components of B and C are combined with the zeroth order (diagonal) component of $\mathsf{D}^{-1}(-\mathsf{B}^{(1)}(\mathsf{D}^0)^{-}\mathsf{C}^{(1)})$. The explicit expressions for B, C, A, and D given in Eqs. (16–22) allow these second-order components to be evaluated in a straightforward manner.

The third-order diagrams in Fig. 2 labeled A1–A6 arise in Eq. (13) from the third-order components of A, with diagrams A1 and A2 coming from matrix elements involving the first-order Rayleigh–Schrödinger component of the wave function $|0\rangle$ and diagrams A3–A6 resulting from elements involving the second-order component of $|0\rangle$. Diagrams C1–C6 and D1–D6 arise in Eq. (13) from three sources: (1) by combining the second-order parts of C, the zeroth-order parts of D^{-1}, and the first-order parts of B; (2) by combining the first-order parts of C, the zeroth-order parts of D^{-1}, and the second-order parts of B; and (3) by combining the first-order parts of B and C with the first-order parts of D^{-1}. Because D contains zeroth- and first-order components $D^0 + D^1$ [see Eqs. (19) and (20)], D^{-1} can be approximated, through first order, and $(D^0)^{-1} - (D^0)^{-1}D^1(D^0)^{-1}$; the second term in this expression is that referred to above as the first-order part of D^{-1}.

In solving the Dyson equation (23) for $G(E)$, Cederbaum must evaluate the contribution to $\sum_{ij}(E)$ made by each diagram in Figs. 1 and 2 for each

Fig. 1. The second-order self-energy diagrams of Cederbaum.

value of the variable E. The evaluation of the third-order diagrams C1–C6 and D1–D6 involves the summation over five indices, e.g.,

$$C1 = \sum_{\substack{\alpha \\ p<q,\, r<s}} \langle pq| |i\alpha\rangle\langle rs| |pq\rangle\langle j\alpha| |rs\rangle \times (\varepsilon_\alpha - \varepsilon_p - \varepsilon_q + E)^{-1}(\varepsilon_\alpha - \varepsilon_r - \varepsilon_s + E)^{-1}. \quad (24)$$

The computation time involved in calculating such terms clearly depends upon the number of Hartree–Fock spin orbitals to the fifth power. This N^5 dependence is obviously a severely limiting aspect of the Green's function method as implemented in the above manner. Nevertheless, Cederbaum (1973) and his collaborators (1971, 1973) have carried out high precision

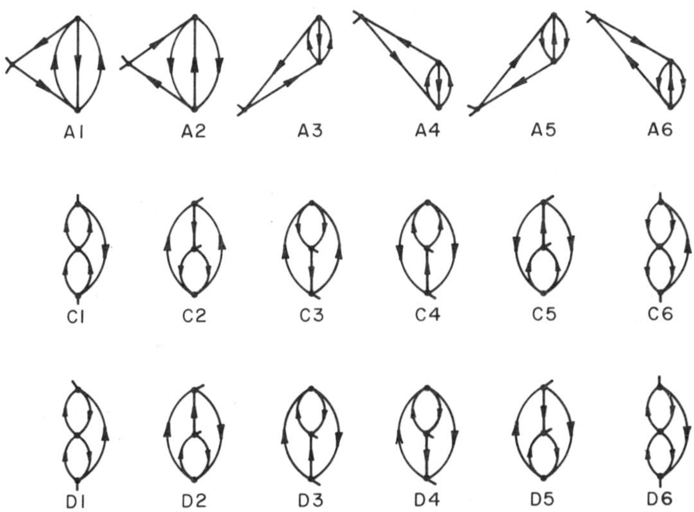

Fig. 2. The third-order self-energy diagrams of Cederbaum.

studies of reasonably large molecules using this complete third-order treatment of $\Sigma(E)$.

B. Simons' Approach and the Technique of Purvis and Öhrn

To accomplish a reduction in the computational effort required to compute $\Sigma(E)$ through third order, Simons and Smith (1973) introduced, in an *ad hoc* manner, simplifications of $\Sigma(E)$ that have now been wisely used in their studies of molecular anions (Smith *et al.*, 1974; Griffing and Simons, 1975, 1976; Kenney and Simons, 1975; Andersen and Simons, 1976; Griffing *et al.*, 1975; Jordan *et al.*, 1976a; Andersen and Simons, 1977a). These authors approximated the D matrix appearing in Eq. (13) by its diagonal component only; all off-diagonal components of D contribute to the third-order parts of $\Sigma(E)$ through terms of the form $-\mathsf{B}^{(1)}(\mathsf{D}^0)^{-1}\mathsf{D}^1(\mathsf{D}^0)^{-1}\mathsf{C}^1$, therefore this approximation causes errors in $\Sigma(E)$ which are of third order. Purvis and Öhrn (1974, 1975) and P. Jørgensen and G. Purvis (private communication) have been involved in attempting to develop more systematic approximations to D^{-1}. Their emphasis has been directed toward finding computationally tractable schemes that allow one to replace the full D matrix by a smaller sub-block of D. Thus far there have not been many calculations carried out using such block diagonalization techniques, although their implementation is certainly worth pursuing.

In the approaches of Simons *et al.* and of Purvis and Öhrn, the self-energy $\Sigma(E)$ contains, as it should, only simple poles; only D is E-dependent. Thus the first- and second-order components of B and C, which are not E-dependent, need be calculated only once. Only D^{-1} has to be formed for each value of E. If D is taken, as Simons has done, to be diagonal, this is a trivial step. In contrast, as is shown in Eq. (24), the diagrammatic representation of $\Sigma(E)$, although very closely related to the algebraic superoperator approach given in Section II, requires each contribution to $\Sigma(E)$ to be calculated for each value of E. Thus the more rigorous treatment of Cederbaum involves carrying out fivefold summations for each value of E and for each element $\sum_{ij}(E)$ of the self energy [see Eq. (24), for example]. If D is taken to be diagonal, the approach described in Section II involves doing a threefold summation for each element of $\Sigma(E)$, once the E-independent quantities A, B, and C are computed (once-and-for-all). The replacement of fivefold sums by threefold sums represents an important savings.

C. Chong's Third-Order Perturbation Theory Approach

In a recent series of formal and computational papers, Chong *et al.* (1974a,b,c) has employed third-order Rayleigh–Schrödinger perturbation theory with geometric approximations to higher order terms and double-zeta quality basis sets to study the low-lying states of molecular ions (primarily cations). This approach involves the separate calculation, via third-order perturbation theory, of the energies of both the neutral molecule and the ion. Such perturbation calculations have yielded ionization potentials that are precise to ± 0.5 eV. In studying anions, this level of accuracy is not adequate. It is our feeling that much of the error resulting in Chong's calculations is due to the quality of his basis sets. A critical evaluation of this third-order perturbation theory approach in which extended basis sets are employed would be most welcome. Once the basis set question is resolved, one can attempt to evaluate the inherent limitations of the perturbation theory method.

D. The Ionization Energy Methods of Parr and of Smith

Parr (Morrell *et al.*, 1975) and Smith (Smith and Day, 1974; Day *et al.*, 1975) have independently developed a technique for computing ionization potentials that is closely related to the Green's function approach described in Section II. The implementation of this theory involves construction of a matrix V whose elements are given (for IP calculations) by

$$V_{ij} = \tfrac{1}{2}\langle 0 | a_i^+ [H, a_j] + [a_j^+, H] a_i | 0 \rangle, \tag{25}$$

where the reference state $|0\rangle$ is assumed to be a high quality (correlated) approximation to the ground electronic state of the system. The (approximate) ionization energies $\{\varepsilon_\mu\}$ are obtained by solving

$$\mathsf{V} g_\mu = \varepsilon_\mu \mathsf{S} g_\mu, \tag{26}$$

with

$$S_{ij} \equiv \langle 0 | a_i^+ a_j | 0 \rangle. \tag{27}$$

This approach has been applied to a few atomic systems; the results are sufficiently encouraging to warrant further investigation of the idea. Andersen (Andersen and Simons, 1976) has carried out an order-by-order analysis of the elements of V by using a second-order Rayleigh–Schrödinger approximation to the reference state in Eq. (25). He finds that the (primary) ionization energies generated by the method of Parr and Smith contain almost all second- and third-order corrections to the Koopmans' theorem approximation, if the second-order reference state is used for $|0\rangle$ in Eq. (25). The method therefore seems to also have a good deal of formal justification. However, it is not at all clear how well this approach will do when it is applied to shake-up ionization processes. It is our feeling that formal and computational investigations of this question must be accomplished before a complete and fair evaluation of the method of Parr and Smith can be made.

IV. The Future

The most important problems and unanswered questions to be addressed in the near future are as follows.

1. How can one optimally treat shake-up ionization? In relation to studies of anions, this problem has to do with treating Feshbach resonance states in which an electron binds to a molecule in an excited state. Both the energies (resonance positions) and lifetimes (transition probabilities) are of utmost importance.

2. Can the (very large) D^{-1} matrix of Eq. (13) be treated in an approximate and computationally tractable, but reasonably accurate, manner? The full calculation of $\sum_{ij}(E)$ for all values of E is probably too large a task for molecules with more than 30 valence electrons. Techniques aimed at "imaging" $\sum_{ij}(E)$ in terms of a small number of parameters must be pursued actively.

3. The development of Green's function techniques capable of studying the shape resonances of metastable anions is important. As in the case of Feshbach resonances, the positions and lifetimes of the states are the relevant quantities. The marriage of Green's function methods and the stabilization technique employed by Taylor (1970) looks like a promising approach to this problem.

4. The treatment of vibrational effects in the photodetachment spectra of molecular anions needs a great deal of work. Cederbaum and Domke (1976) have made nice progress in dealing with vibrational structure, including electronic–nuclear coupling, of molecular photoelectron spectra. The application of the kind of approach used by Cederbaum to the problem of photodetachment of anions could yield important new information.

References

Andersen, E., and Simons, J. (1976). *J. Chem. Phys.* **64**, 4548–4550.
Andersen, E., and Simons, J. (1977a). *J. Chem. Phys.* **65**, 5393–5397.
Andersen, E., and Simons, J. (1977b). *J. Chem. Phys.* **66**, 2427–2430.
Berry, R. S. (1969). *Chem. Rev.* **69**, 533–542.
Branscomb, L. M. (1966). *Phys. Rev.* **148**, 11–18.
Cade, P. E. (1967a). *J. Chem. Phys.* **47**, 2390–2406.
Cade, P. E. (1967b). *Proc. Phys. Soc., London* **91**, 842–854.
Cederbaum, L. S. (1973). *Theor. Chim. Acta* **31**, 239–260.
Cederbaum, L. S., and Domke, W. (1976). *J. Chem. Phys.* **60**, 2878–2889.
Cederbaum, L. S., Holneicher, G., and Peyerimhoff, S. (1971). *Chem. Phys. Lett.* **11**, 421–424.

Cederbaum, L. S., Holneicher, G., and von Niessin, W. (1973). *Chem. Phys. Lett.* **18**, 503–508.
Chong, D. P., Herring, F. G., and McWilliams, D. (1974a). *Chem. Phys. Lett.* **25**, 568–570.
Chong, D. P., Herring, F. G., and McWilliams, D. (1974b). *J. Chem. Phys.* **61**, 3567–3570.
Chong, D. P., Herring, F. G., and McWilliams, D. (1974c). *J. Chem. Soc., Faraday Trans. 2* **70**, 193–202.
Clementi, E. (1964). *Phys. Rev.* **135**, A980–A984.
Clementi, E., and McLean, A. D. (1964). *Phys. Rev.* **133**, A419–A423.
Clementi, E., McLean, A. D., Raimondi, D. L., and Yoshimine, M. (1964). *Phys. Rev.* **133**, A1274–A1279.
Day, O. W., Smith, D. W., and Morrison, R. C. (1975). *J. Chem. Phys.* **62**, 115–119.
Doll, J. D., and Reinhardt, W. P. (1972). *J. Chem. Phys.* **57**, 1169–1184.
Goscinski, O., and Lukman, B. (1970). *Chem. Phys. Lett.* **7**, 573–576.
Griffing, K., and Simons, J. (1975). *J. Chem. Phys.* **62**, 535–540.
Griffing, K., and Simons, J. (1976). *J. Chem. Phys.* **64**, 3610–3614.
Griffing, K., Kenney, J., Simons, J., and Jordan, K. D. (1975). *J. Chem. Phys.* **63**, 4073–4075.
Hotop, H., Patterson, T. A., and Lineberger, W. C. (1974). *J. Chem. Phys.* **60**, 1806–1812.
Jordan, K. D., Griffing, K. M., Kenney, J., Andersen, E. L., and Simons, J. (1976a). *J. Chem. Phys.* **64**, 4730–4740.
Jordan, K. D., Chen, T. T., and Simons, J. (1976b). *Chem. Phys.* **14**, 145–158.
Jørgensen, P., and Simons, J. (1975). *J. Chem. Phys.* **63**, 5302–5304.
Kenney, J., and Simons, J. (1975). *J. Chem. Phys.* **62**, 592–599.
Linderberg, J., and Öhrn, Y. (1967). *Chem. Phys. Lett.* **1**, 295–296.
Linderberg, J., and Öhrn, Y. (1973). "Propagators in Quantum Chemistry." Academic Press, New York.
Morrell, M. M., Parr, R. G., and Levy, M. (1975). *J. Chem. Phys.* **62**, 549–554.
Pekeris, C. L. (1958). *Phys. Rev.* **112**, 1649–1658.
Pickup, B. T., and Goscinski, O. (1973). *Mol. Phys.* **26**, 1013–1035.
Purvis, G., and Öhrn, Y. (1974). *J. Chem. Phys.* **60**, 4063–4069.
Purvis, G., and Öhrn, Y. (1975). *J. Chem. Phys.* **62**, 2045–2049.
Purvis, G., Nehrkorn, C., and Öhrn, Y. (1976). *J. Chem. Phys.* **64**, 1752–1756.
Schneider, B., Taylor, H. S., and Yaris, R. (1970). *Phys. Rev. A* **1**, 855–867.
Simons, J. (1974). *Chem. Phys. Lett.* **25**, 122–128.
Simons, J. (1977). *Annu. Rev. Phys. Chem.* (in press).
Simons, J., and Smith, W. D. (1973). *J. Chem. Phys.* **58**, 4899–4907.
Sinanoğlu, O., and Oksuz, I. (1968). *Phys. Rev. Lett.* **21**, 507.
Smith, D. W., and Day, O. W. (1974). *J. Chem. Phys.* **62**, 113–114.
Smith, W. D., Chen, T. T., and Simons, J. (1974). *Chem. Phys. Lett.* **27**, 499–502.
Taylor, H. S. (1970). *Adv. Chem. Phys.* **18**, 91–113.
Taylor, H. S., and Harris, F. E. (1963). *J. Chem. Phys.* **39**, 1012–1016.
Tsui, F., and Freed, K. F. (1975). *Chem. Phys. Lett.* **32**, 345–350.
Tyner-Redmon, L., Purvis, G., and Öhrn, Y. (1975). *J. Chem. Phys.* **63**, 5011–5017.
Wahl, A. C., and Gilbert, T. L. (1965). *Bull. Am. Phys. Soc.* [2] **10**, 1097.
Weiss, A. W. (1968). *Phys. Rev.* **166**, 70–73.
Yarlagadda, B. S., Csanak, Gy., Taylor, H. S., Schneider, B., and Yaris, R. (1973). *Phys. Rev. A* **7**, 146–153.
Zemke, W. T., Das, G., and Wahl, A. C. (1972). *Chem. Phys. Lett.* **14**, 310–314.

Geometrical Aspects of Equilibrium Thermodynamics

Frank Weinhold

Theoretical Chemistry Institute and Department of Chemistry
University of Wisconsin–Madison, Madison, Wisconsin

> Although geometrical representations of propositions in the thermodynamics of fluids are in general use, and have done good service in disseminating clear notions in this science, yet they have by no means received the extension in respect to variety and generality of which they are capable.
>
> J. W. Gibbs (1873a)

I. Introduction

From its inception, the historical development of thermodynamics has been curiously interwoven with the mathematics of geometry. Sadi Carnot, who founded thermodynamics, was educated to age sixteen by his father Lazare, a distinguished mathematician who wrote on abstract geometry and principles of mechanical equilibrium (Mendoza, 1960). Hermann L. von Helmholtz, who gave the first comprehensive statement of the first law of thermodynamics, also made important mathematical contributions to the theory of non-Euclidean geometry, worthy to be mentioned alongside the work of his contemporaries, Riemann and Lobachevsky (Russell, 1956). J. Willard Gibbs, who raised thermodynamic theory to an awesome perfection, established the wide applicability of geometric methods in thermodynamic analysis, and later made important contributions to geometric theory through his vector algebra—the notation and techniques of which are still widely employed.

But the association of thermodynamics and geometry, in whatever form, would seem unlikely in at least two senses. First, geometry is among the purest examples of deductive and mathematical inquiry, while thermodynamics was historically an inductive and experimental science, rooted in engineering applications, and resting on laws that had emerged only haltingly from empirical evidence. Thus, the basic *methodologies* of the two subjects seem to be entirely dissimilar. Second, geometry was historically concerned with the spatial sizes, shapes, and orientations of objects in ordinary space. By contrast, thermodynamics (perhaps uniquely among the natural sciences) is concerned with systems whose sizes, shapes, and orientations are regarded as *un*important; only such internal properties as temperature and pressure serve to distinguish one equilibrium system from another. Thus, the *objects* of a thermodynamic theory seem quite the opposite of those encompassed by geometry.

Nevertheless, from the time of Gibbs it has been realized that the relationship between thermodynamics and geometry is very strong indeed, and many of the recent advances in thermodynamic theory are based directly on the geometrical aspects of that theory as recognized by Gibbs (Mountain, 1974). More recently (Weinhold, 1975a-d, 1976a-c) it was recognized that the relationship between thermodynamics and geometry goes considerably beyond that previously envisioned—extending to fundamental structural features of the two theories and finally unifying them in a common mathematical framework. This observation has both conceptual and practical implications for thermodynamic theory, yet it has seemed preferable (in a serial volume devoted to "perspectives") to stress the former aspects in the present article.

Though the metric geometry to be discussed below is quite distinct from that introduced into thermodynamics by Gibbs, its origins lie directly in the Gibbs formalism, which in turn had evolved from considerations which were "geometrical" in a quite different sense. The present article therefore properly begins with a brief account of the thermodynamic papers of Gibbs, and a mathematical discussion of what is implied by various usages of the term "geometry." It was not feasible within reasonable constraints of length to include details of the metric formalism itself or its application to specific problems (see Section III,D). Rather, I have concentrated on laying out physical and mathematical foundations of this formalism in somewhat greater detail than was given in the original presentation (Weinhold, 1975a), and in incorporating elements needed for the treatment of heterogeneous equilibrium (Weinhold, 1976c). Since several specific applications of the geometrical formalism have been described in the primary literature (Weinhold, 1975b-d, 1976c), the present emphasis on *where it comes from*, rather than *what it does*, may be adequately motivated.

In Section IV, discussion is given of some related geometrical aspects of thermodynamic theory. These include the formalisms of Baur *et al.* (1965) and of Tisza (1966), the possible relationship between Gibbs' phase rule and Euler's topological formula, the "thermodynamic square" and associated geometrical mnemonics, and other miscellaneous topics.

Concluding remarks are contained in Section V.

II. Graphical and Geometrical Methods in Gibbs' Thermodynamics

A. Gibbs' First Paper: Graphical Methods in the Thermodynamics of Fluids†

Graphical methods for representing the relationships among thermodynamic properties antedate even the earliest steps toward a coherent theory of thermodynamic processes. The Scottish engineer James Watt, inventor of the modern condensing steam engine, made use of pressure–volume diagrams in his design studies, and later invented a mechanical device, Watt's indicator, which automatically recorded the PV diagram for the compressed steam as the piston stroke progressed. This instrument, it has been said, "is to the steam engineer what the stethoscope is to the physician."

† Gibbs' thermodynamic papers are reprinted in Vol. I of the *Collected Works* (Gibbs, 1928), to which the page numbers cited below will refer. The *Commentary* on Gibbs' works (Donnan and Haas, 1936) includes extensive discussion of the third paper (Gibbs, 1876–1878), but little concerning the formative first and second papers (Gibbs, 1873a,b).

By the time of Gibbs' first published work, "Graphical Methods in the Thermodynamics of Fluids" (Gibbs, 1873a), the two-dimensional pressure–volume diagrams were in general use. The use of a three-dimensional surface to represent simultaneous values of pressure, volume, and temperature had also been previously proposed by James Thomson (1871, 1872), the brother of William Thomson, Lord Kelvin. Yet Gibbs recognized such entirely new possibilities in the geometrical and graphical methods that their use is generally associated with his name. Gibbs' first paper is particularly interesting in suggesting how he came to see in the geometrical features of thermodynamic graphs an important source of insights concerning thermodynamic theory itself. Already in this first work one can recognize the thoroughness, the accurate logic, and the natural tendency toward elegance and conciseness of mathematical method that marked his writings.

After remarking on the "good service" done by geometrical representations of thermodynamic propositions, and the near-exclusive use of the pressure–volume diagram by other workers, Gibbs states it as his objective "to call attention to certain diagrams of different construction, which afford graphical methods coextensive in their applications with that in ordinary use, and preferable to it in many cases in respect of distinctness or of convenience." He then sets out the relations that are to be represented by such diagrams, the properties that all such diagrams must possess, and his criterion for selecting the most suitable diagram, viz., "consideration of simplicity and convenience."

The scope of this inquiry might initially seem quite restrictive, and perhaps to concern mere matters of pedagogical detail,† yet Gibbs derives thermodynamic insights of very general utility. Perhaps it was simply the observation that two thermodynamic diagrams are often not of *equivalent* convenience which led him to seek the *most* convenient type of diagram, and to see in such graphical convenience an active guide to general thermodynamic relationships. Such a viewpoint strongly impelled his later work on vector analysis and multiple algebra, in which he stressed the importance of identifying the mathematical formulation for which the most important relationships could be most simply and naturally expressed.

Gibbs then discusses the properties of various diagrams in some detail, including—in addition to the now-familiar PV, ST, and VS diagrams—such possibilities as log V-log P, S-log T, S-log V, etc. It seems likely that Gibbs

† That Gibbs was attentive to matters of pedagogy may be indicated by his remark (p. 11) concerning the ST diagram: "If, then, it is more important for purposes of instruction and the like to familiarize the learner with the second law, than to defer its statement as long as possible, the use of the entropy–temperature diagram may serve a useful purpose in the popularizing of this science."

was particularly struck by the simplicity of the ideal Carnot cycle in the ST diagram, with its easy representation of the work carried out and the heat evolved. Yet most attention was drawn to the volume–entropy diagram, for its clarity in representing properties of coexisting states and heterogeneous equilibrium. Because the axes are both proportional to the quantity of substance, the VS diagram is useful when the state properties of a substance, rather than the work or heat of some cycle, are of primary interest. The advantages of the VS diagram might be anticipated, as Gibbs remarks, "from the simple and symmetrical form of the general equations of thermodynamics, when volume and entropy are chosen as independent variables." Such considerations of course led Gibbs subsequently to the USV surface, whose properties were so completely developed in the second paper.

Several points are worth noting concerning Gibbs' first paper on graphical methods.

1. Gibbs clearly noted that the association of thermodynamic states with points on a two-dimensional plane supposes nothing more than a certain condition of continuity and is thus devoid of metric significance. As he states, "The different diagrams which we obtain by different laws of association are all such as may be obtained from one another by a process of *distortion*," and the use of rectilinear coordinates is therefore an arbitrary (if convenient) element of such graphical analyses.

2. Gibbs recognized that the possibility of using plane diagrams to represent the thermodynamics of fluids "evidently arises from the fact that the state of the body considered, like the position of a point in a plane, is capable of two and only two independent variations." One sees in these words an anticipation of the reasoning that culminated in Gibbs' phase rule, and explicit recognition that the dimensionality of the diagram is connected with the number of degrees of freedom which is required (on empirical grounds) to specify the thermodynamic state.

3. Gibbs recognized the "independence and sufficiency of a graphical method," i.e., that the diagrams might be taken to *replace*, rather than merely to illustrate, the analytical formulas by which thermodynamic principles were usually expressed. Consideration of the geometry of these diagrams was therefore able to stand alongside analytical manipulations of thermodynamic formulas as a source of valid thermodynamic conclusions.

It is interesting to note the great stress Gibbs placed on the *entropy* concept (because of its role in making the second law appear in a particularly simple form) at a time when the subject still provoked much confusion (see footnote, p. 51). Thus, while arguing the virtue of the *ST* diagram, Gibbs found it necessary to apologize for invoking a quantity whose "very existence . . . will doubtless seem to many far-fetched, and may repel beginners as obscure and difficult of comprehension."

B. Gibbs' Second Paper: A Method of Geometrical Representation of the Thermodynamic Properties of Substances by Means of Surfaces

In his second paper, Gibbs (1873b) extended the thermodynamic diagrams into three dimensions. Noting that the thermodynamic properties of a fluid are determined by the relations between P, V, T, U, and S (for fixed mass), Gibbs proposes to represent these relations by a thermodynamic surface whose points are made equal to the energy, entropy, and volume of the fluid in a three-dimensional Cartesian frame. Gibbs points out the superiority of this USV surface over the PVT surface used previously by Thomson, since differentiation of $U = U(S, V)$ at any point of the surface indicates the temperature and pressure by the relations

$$T = (\partial U/\partial S)_V, \qquad -P = (\partial U/\partial V)_S$$

and thus permits construction of the PVT surface, whereas the latter surface does not suffice to construct the former. The abstract geometry of this three-dimensional thermodynamic surface gave ample rein to Gibbs' unusual capacities for geometrical visualization.

Gibbs carefully distinguishes at the outset "between what is essential and what is arbitrary in a surface thus formed." By considering tangent planes on this surface, he shows how a "derived" surface can be formed as the locus of lines connecting points of contact of a double-tangent plane rolling over the "primitive" surface, such rolling being physically permitted by the convexity of the primitive surface (which is in turn a consequence of thermodynamic stability). Such derived surfaces are shown to correspond to the coexistent phases of heterogeneous equilibrium, so that a geometrical analysis of phase equilibria becomes possible. This is carried out with characteristic thoroughness, as Gibbs describes the special geometrical features associated with coexistence lines, the triple point, and (particularly) the fluid critical point, where the lines of double tangency merge to a common terminus. So useful were these geometrical relationships in clarifying phase equilibria that Maxwell (1875) devoted thirteen pages to their exposition in the fourth edition of his *Theory of Heat*, and personally sent Gibbs a plaster model of the thermodynamic surface of water which the latter is said to have used regularly in his lectures on thermodynamics at Yale (Wilson, 1936).

C. Gibbs' Third Paper: On the Equilibrium of Heterogeneous Substances

With the completion of the second paper, Gibbs' graphical methods may be said to have reached their zenith, for in the great third paper Gibbs

(1876–1878) turned his attention in quite another direction. Whereas the second paper was very largely concerned with geometrical properties of global thermodynamic surfaces, the third paper makes no reference to such surfaces for some 60 pages (and thereafter employs them, in a section entitled "Geometrical Illustrations," only to illustrate results that were obtained elsewhere by analytic methods). Of course, a principal reason for this abrupt shift is Gibbs' desire to incorporate chemical systems of several components, for as he notes (p. 362), "when the composition of the body is variable, the fundamental equation cannot be completely represented by any surface or finite number of surfaces." Such many-component systems therefore defy the three-dimensional sense of visualization which Gibbs was able to exploit in the case of simple fluids.

Yet there is a deeper sense in which Gibbs turns away from such surfaces, with their global inclusion of the totality of equilibrium states accessible to a system. At the beginning of the third paper he sets out the "criteria of equilibrium and stability" in a form which makes reference only to properties of one particular equilibrium state, the essential content of the first and second laws of thermodynamics thereby being given a local form which makes no reference to engines, cycles, perpetual-motion devices, and the like. By this step, Gibbs makes the equilibrium states, rather than processes, the basic conceptual elements of his thermodynamic theory (Tisza, 1966, p. 40).

In this change of perspective from the global to the local, the fundamental equation $U = U(S, V, N_1, N_2, \ldots)$ may be thought of as providing a parametrization of the equilibrium states in "Gibbs space" (Tisza, 1966, pp. 49–51), and the properties of any one such state are developed by analytical methods involving at most the lowest few (i.e., first and second) derivatives of this function. Thus, in Gibbs' treatment it is the thermodynamic relations among properties (for some fixed state) rather than the relations connecting different states that come to be emphasized. For the purpose of deriving such relations, graphical methods of Gibbs space prove less convenient than more conventional analytical methods.

The wish to treat complex many-component systems thus apparently limited the useful role that geometry could play in a thermodynamic theory. Yet, paradoxically, it is precisely for such systems that a metric geometrical theory of equilibrium thermodynamics becomes most advantageous.

III. Metric Geometry of Equilibrium Thermodynamics

A. Abstract Metric Geometries

In studying the properties of a physical or mathematical system, one can sometimes recognize that the results obtained depend not so much on the precise definition of the system as on the formal relations holding among its objects. In such a case, this relational structure—the rules of manipulation of the objects, or briefly, their algebra—may itself be adopted as the object of study, so that the properties of the original system will be seen as but a special instance of that structure. Because the empirical laws of thermodynamics furnish constraints on the relations among equilibrium systems and their properties, a study of such abstract relational structures will be naturally suggested.†

1. Linear Manifolds and Metric Spaces

Among the simplest algebraic structures which can characterize a set of objects $\{\mathscr{X}, \mathscr{Y}, \mathscr{Z}, \ldots\}$ is that of a *linear manifold* (also called a linear vector space, affine space, etc.). By definition, such a manifold has only the two operations of addition $(\mathscr{X} + \mathscr{Y})$ and multiplication by a scalar $(\lambda\mathscr{X})$, the result in each case to give an element of the manifold. These operations furthermore have the usual (distributive, associative, and commutative) properties which their names imply. A linear manifold therefore possesses the qualities of connectivity or continuity implied by the word "space," but lacks the qualities of measurability, e.g., the possibility of assigning numbers to each pair of objects which could reflect their "proximity." The geometry of such a linear manifold therefore has only an affine or topological (rather than a metric) significance, and cannot be associated with geometry in its historical Euclidean sense as applying to the distances, angles, areas, and other metric characteristics of spatial objects. Gibbs' geometrical methods, being based on thermodynamic surfaces that are subject to a process of distortion, therefore have only an affine (rather than a metric, or Euclidean) significance.

In order to gain the additional geometrical structure of a *metric space*, one requires in addition a *scalar product* $\langle\mathscr{X}|\mathscr{Y}\rangle$, i.e., a rule which associates a (real) scalar with every pair of objects $\mathscr{X}$, $\mathscr{Y}$ in the manifold. In ordinary Euclidean vector geometry, this product is obtained by multiplying the lengths of two vectors by the cosine of their angle of separation, and can readily be shown to have the properties

$$\text{distributive:} \quad \langle\mathscr{X}|\lambda\mathscr{Y} + \mu\mathscr{Z}\rangle = \lambda\langle\mathscr{X}|\mathscr{Y}\rangle + \mu\langle\mathscr{X}|\mathscr{Z}\rangle \tag{3.1a}$$

$$\text{symmetric:} \quad \langle\mathscr{X}|\mathscr{Y}\rangle = \langle\mathscr{Y}|\mathscr{X}\rangle \tag{3.1b}$$

$$\text{positive:} \quad \langle\mathscr{X}|\mathscr{X}\rangle \geq 0 \ (= 0 \text{ only if } \mathscr{X} = 0). \tag{3.1c}$$

However, it is a remarkable fact that these three properties of the Euclidean

† For general mathematical background on the material discussed in this section, see, e.g., Dennery and Krzywicki (1967), Shilov (1961), Mirsky (1955), and Misner *et al.* (1973).

scalar product are not only necessary consequences of the Euclidean structure, but are also fully sufficient to insure a complete Euclidean geometry. In other words, if any rule can be found which associates a number $\langle \mathscr{X} | \mathscr{Y} \rangle$ with the objects $\mathscr{X}$, $\mathscr{Y}$ of an abstract linear manifold in a way which satisfies conditions (3.1), then the manifold is isomorphic to a corresponding Euclidean vector space (see, e.g., Dennery and Krzywicki, 1967). The axiomatic requirements (3.1) can be shown immediately to imply the Schwarz inequality

$$\langle \mathscr{X} | \mathscr{Y} \rangle^2 \leq \langle \mathscr{X} | \mathscr{X} \rangle \langle \mathscr{Y} | \mathscr{Y} \rangle \tag{3.2}$$

for every pair of objects $\mathscr{X}$, $\mathscr{Y}$ of the metric space. Moreover, requirement (3.1c) permits one to consistently define the (positive) "distance" $|\mathscr{X} - \mathscr{Y}|$ between objects $\mathscr{X}$ and $\mathscr{Y}$ as

$$|\mathscr{X} - \mathscr{Y}| \equiv \langle \mathscr{X} - \mathscr{Y} | \mathscr{X} - \mathscr{Y} \rangle^{1/2}. \tag{3.3a}$$

Distance thus defined, like its Euclidean counterpart, will be found to satisfy the triangle inequality

$$|\mathscr{X} - \mathscr{Y}| \leq |\mathscr{X} - \mathscr{Z}| + |\mathscr{Y} - \mathscr{Z}|. \tag{3.3b}$$

Thus, while the objects of the metric space may initially have arisen with no reference to geometry, they can be labelled as vectors and manipulated by the familiar rules of Euclidean geometry, for they are in one-to-one correspondence with the elements of a Euclidean vector space, and their behavior (mathematical relationships, rules of combination, etc.) is in all respects identical with the behavior of corresponding Euclidean vectors.

2. *Dimensionality and Metric*

The abstract metric spaces $\mathfrak{M}$, as defined axiomatically through Eqs. (3.1), may be distinguished by their dimensionality r, a fundamental property which can be determined "experimentally" from numerical values of the scalar products. Suppose, for example, that k vectors $\{\mathscr{Y}^{(k)}\} = \{\mathscr{Y}_1, \mathscr{Y}_2, \ldots, \mathscr{Y}_k\}$ of $\mathfrak{M}$ have been found for which

$$\det |\mathsf{G}^{(k)}| \neq 0 \tag{3.4}$$

where $\mathsf{G}^{(k)} = \mathsf{G}(\{\mathscr{Y}^{(k)}\})$ is the matrix of scalar products among the $\mathscr{Y}_i$'s,

$$(\mathsf{G}^{(k)})_{ij} \equiv \langle \mathscr{Y}_i | \mathscr{Y}_j \rangle, \qquad i, j = 1, 2, \ldots, k. \tag{3.5}$$

The associated *conjugate* vectors $\overline{\mathscr{Y}}_i$ satisfying

$$\langle \mathscr{Y}_i | \overline{\mathscr{Y}}_j \rangle = \delta_{ij}, \qquad i, j = 1, 2, \ldots, k \tag{3.6}$$

allow one to construct the projection operator $\mathscr{O}^{(k)}$,

$$\mathscr{O}^{(k)} \equiv 1 - \sum_{i=1}^{k} |\mathscr{Y}_i\rangle\langle \overline{\mathscr{Y}}_i| \tag{3.7}$$

which, when applied to each element of $\mathfrak{M}$, gives the subspace $\mathfrak{M}^{(k)}$ $(\equiv \mathcal{O}^{(k)}\mathfrak{M})$ which is the orthogonal complement of that spanned by the set $\{\mathcal{Y}^{(k)}\}$. Now if $\mathfrak{M}^{(k)}$ contains any nonzero element, that element may be chosen as a new $\mathcal{Y}_{k+1}$, for it is easy to show that the resulting $\det|\mathsf{G}^{(k+1)}|$ is nonzero. The cycle through Eqs. (3.4)–(3.7) can then be repeated with the new $\mathcal{O}^{(k+1)}$ and $\mathfrak{M}^{(k+1)}$, the latter being examined for nonzero elements. If, for any k, the subspace $\mathfrak{M}^{(k)}$ has only vectors of zero length, then $k = r$, and the dimensionality of the full space has been determined. Put more simply, the dimensionality r is that largest integer for which

$$\det|\mathsf{G}^{(r)}| \neq 0 \tag{3.8}$$

for some choice of r vectors, but

$$\det|\mathsf{G}^{(r+1)}| = 0 \tag{3.9}$$

for every choice of $r + 1$ vectors. Conversely, knowledge of the dimensionality r allows useful constraints [of the form of Eqs. (3.8) or (3.9)] to be put on possible numerical values of the scalar products. We label the metric space $\mathfrak{M}_r$ to explicitly exhibit its dimensionality.

A set of r vectors $\{\mathcal{Y}_1, \mathcal{Y}_2, \ldots, \mathcal{Y}_r\}$ obtained by the procedure of the previous paragraph will form a basis in $\mathfrak{M}_r$. The associated metric matrix $\mathsf{G} = \mathsf{G}^{(r)}$ [cf. Eq. (3.5)]

$$(\mathsf{G})_{ij} \equiv \langle \mathcal{Y}_i | \mathcal{Y}_j \rangle, \qquad i, j = 1, 2, \ldots, r \tag{3.10}$$

$$\det|\mathsf{G}| \neq 0 \tag{3.11}$$

is a fundamental mathematical object of $\mathfrak{M}_r$, determining all characteristic geometric relationships in the space. Quite generally, the coordinate-free metric $\mathcal{G}$ of the space can be regarded as the abstract bilinear functional of arbitrary vectors $\mathcal{U}$, $\mathcal{V}$ which determines their scalar product,

$$\mathcal{G}(\mathcal{U}, \mathcal{V}) \equiv \langle \mathcal{U} | \mathcal{V} \rangle, \tag{3.12}$$

while G is its r-dimensional matrix representation in a particular basis set $\{\mathcal{Y}\}$. Thus, if $\mathcal{U}$ and $\mathcal{V}$ are represented in this basis as

$$\mathcal{U} = \sum_{i=1}^{r} u_i \mathcal{Y}_i \tag{3.13a}$$

$$\mathcal{V} = \sum_{i=1}^{r} v_i \mathcal{Y}_i \tag{3.13b}$$

the metric functional (3.12) takes the form

$$\mathcal{G}(\mathcal{U}, \mathcal{V}) = \mathbf{u}^t \mathsf{G} \mathbf{v}, \tag{3.14}$$

where $\mathbf{u}$, $\mathbf{v}$ are r-dimensional column vectors of the coefficients in (3.13),

$$\mathbf{u} \equiv \begin{bmatrix} u_1 \\ u_2 \\ \vdots \\ u_r \end{bmatrix}, \qquad \mathbf{v} = \begin{bmatrix} v_1 \\ v_2 \\ \vdots \\ v_r \end{bmatrix}. \tag{3.15}$$

Of course, the intrinsic metric $\mathscr{G}$ is more fundamental than any of its matrix representations, which might have been given with respect to many possible choices of basis set. In particular, there is no need to restrict these matrix representations to the minimum dimensionality r. Rather, one might consider an arbitrary set of s vectors (say $s = r + \nu$),

$$\{\mathscr{Y}^{(s)}\} = \{\mathscr{Y}_1, \mathscr{Y}_2, \ldots, \mathscr{Y}_s\} \tag{3.16}$$

for which

$$\mathscr{U} = \sum_{i=1}^{s} (\mathbf{u}^{(s)})_i \mathscr{Y}_i \tag{3.17a}$$

$$\mathscr{V} = \sum_{i=1}^{s} (\mathbf{v}^{(s)})_i \mathscr{Y}_i \tag{3.17b}$$

and

$$(\mathsf{G}^{(s)})_{ij} = \langle \mathscr{Y}_i | \mathscr{Y}_j \rangle, \qquad i, j = 1, 2, \ldots, s. \tag{3.18}$$

The corresponding expression for the metric functional

$$\mathscr{G}(\mathscr{U}, \mathscr{V}) = \mathbf{u}^{(s)t} \mathsf{G}^{(s)} \mathbf{v}^{(s)} \tag{3.19}$$

will be valid so long as $\mathsf{G}^{(s)}$ properly exhibits the intrinsic spatial dimensionality r in the form

$$\operatorname{rank}(\mathsf{G}^{(s)}) = r = \text{dimension}\{\mathfrak{M}_r\}, \tag{3.20}$$

i.e., so long as the chosen basis (3.16) spans $\mathfrak{M}_r$. Such an s-dimensional representation differs from the minimal r-dimensional representation (3.11) in that the expansion coefficients $(\mathbf{u}^{(s)})_i$, $(\mathbf{v}^{(s)})_i$ of Eqs. (3.17) are no longer uniquely defined, and the metric matrix $\mathsf{G}^{(s)}$ is now necessarily singular, having exactly $\nu = s - r$ zero eigenvalues. Thus, it will be possible to find ν linearly independent column vectors $\boldsymbol{\eta}^{(1)}, \boldsymbol{\eta}^{(2)}, \ldots, \boldsymbol{\eta}^{(\nu)}$ for which

$$\mathsf{G}^{(s)} \boldsymbol{\eta}^{(\lambda)} = 0, \qquad \lambda = 1, 2, \ldots, \nu. \tag{3.21}$$

However, the proper metric relationships will still be given uniquely by (3.19) in any basis for which the dimensionality requirement (3.20) is satisfied. As in Eqs. (3.10)–(3.15), we may suppress the superscript s when $s = r$, the representation of minimum dimensionality.

In this context, one should carefully distinguish between two usages of the word "dimensionality"—On the one hand [as in Eq. (3.20)], to denote an intrinsic property (r) of the metric space $\mathfrak{M}_r$, and on the other, to denote merely the number (s) of rows and columns in a set of vectors and matrices chosen to represent the metric functional in equations such as (3.19). The former usage designates a fundamental property of the metric space, the latter only an artifact of its representation.

3. Some Examples

The elementary example of a metric space is the ordinary space of "positions," the local spatial geometry of the physical world. The elements of the manifold are the position vectors $\vec{\mathbf{u}}$ drawn from some fixed origin, and the scalar products are related to the lengths $|\vec{\mathbf{u}}|$, $|\vec{\mathbf{v}}|$ and separation angle θ of these vectors in the familiar way,

$$\mathscr{U} \Leftrightarrow \vec{\mathbf{u}} \tag{3.22a}$$

$$\langle \mathscr{U} | \mathscr{V} \rangle = |\vec{\mathbf{u}}| \cdot |\vec{\mathbf{v}}| \cos \theta. \tag{3.22b}$$

Note that the "$\Leftrightarrow$" in (3.22a) only identifies isomorphic elements without connoting "equality" in a literal sense, whereas (3.22b) is an equation between numbers. The association (3.22) with position vectors of solid geometry leads to the metric space $\mathfrak{M}_3$, while the analogous association with plane geometry leads to $\mathfrak{M}_2$.

Column vectors $\mathbf{u}$ of order r provide the standard example of $\mathfrak{M}_r$, the Euclidean vector space of r dimensions. In this case

$$\mathscr{U} \Leftrightarrow \mathbf{u} = \begin{bmatrix} u_1 \\ u_2 \\ \vdots \\ u_r \end{bmatrix} \tag{3.23a}$$

$$\langle \mathscr{U} | \mathscr{V} \rangle = \sum_{i=1}^{r} u_i v_i = \mathbf{u}^t \mathbf{v}, \tag{3.23b}$$

and the components u_i (taken real) may be identified with projections of the isomorphic position vector $\vec{\mathbf{u}}$ of (3.22a) on a set of Cartesian axes. However, alternative scalar products, based on a positive definite metric matrix $\mathsf{G} \geq 0$, might also be cited

$$\langle \mathscr{U} | \mathscr{V} \rangle = \sum_{i=1}^{r} \sum_{j=1}^{r} u_i G_{ij} v_j = \mathbf{u}^t \mathsf{G} \mathbf{v}, \tag{3.23c}$$

in which the components u_i lack this direct Cartesian significance. In either (3.23b) or (3.23c), the scalar product can be easily recognized to satisfy the axiomatic requirements (3.1).

Another well-known example of a metric space is the Hilbert space of square-integrable wave functions in quantum mechanics. In this case,

$$\mathscr{U} \Leftrightarrow \psi_u(\vec{\mathbf{r}}) \tag{3.24a}$$

$$\langle \mathscr{U} | \mathscr{V} \rangle = \int_{\text{all space}} \psi_u^*(\vec{\mathbf{r}})\psi_v(\vec{\mathbf{r}})\, d^3\vec{\mathbf{r}} \tag{3.24b}$$

and once again the Euclidean properties (3.1) of the scalar product (3.24b) are easily verified. The metric space is in this case of infinite dimensionality, but the isomorphism with (3.23) permits one to derive quantum-mechanical results in either a *matrix*-mechanical representation based on Eqs. (3.23), or an equivalent *wave*-mechanical representation based on Eqs. (3.24), as emphasized by von Neumann and Dirac. The "state vectors" $\psi_u(\vec{\mathbf{r}})$ thus have simple Euclidean characteristics which are frequently exploited; for example, the origin of the "$\leq$" in the mathematical derivation of Heisenberg's uncertainty relations (Dicke and Wittke, 1960) lies ultimately in the Schwarz inequality (3.2), which is essentially the geometrical restriction, $\cos^2 \theta \leq 1$, for the angle between vectors in Hilbert space.

In quantum mechanics, the notion of an abstract metric space retains its validity even when the associated state vectors have no representation [as in (3.24a)] as functions of spatial coordinates. Thus, the abstract spin, isospin, and similar degrees of freedom belong to spaces which are only attributed the abstract metric properties (3.1). The dimensionalities of such spaces are of particular importance, since they are connected with the range of possible quantum numbers which the associated degree of freedom can assume. Note that such abstract metric spaces are not based on any obvious line element, a traditional point of departure for introducing metric spaces in elementary texts.

A still more general example of a metric space is provided by the classical geometry of Riemann. Riemann's formalism makes possible a distinction between the space of vectors whose metrical relationships are specified by the functional $\mathscr{G}$, and an associated linear manifold by which these vectors and the metric $\mathscr{G}$ might be parametrized. Let ξ be an element of a linear manifold (itself having perhaps no metric character whatsoever) which can parametrize the positions (or, more generally, the states) of a collection of metrical objects $\{\mathscr{Y}\}$. The Riemann geometry permits the metric $\mathscr{G}$ to itself be a function of state,

$$\mathscr{G} = \mathscr{G}_\xi \tag{3.25}$$

so that the geometry of the objects $\{\mathscr{Y}\}$ changes with changes in ξ. For any particular ξ, the Riemann metric satisfies the usual conditions (3.1) for a metric space, but one now considers the multiply continuous family of such spaces, related through the functional dependence of the metric $\mathscr{G}$ on the state ξ.

In the application of Riemann's formalism to geometry, the manifold ξ is identified with a set of positions in a global space, while the metrical relationships (governed by $\mathscr{G}_\xi$) concern infinitesimal displacements in the same space. However, it has proved useful, as in general relativity theory (Misner *et al.*, 1973), to regard metric as depending on qualities ξ (e.g., the masses of physical objects) quite unrelated to the spatial properties which the geometry was intended to describe.

In a similar vein, the Riemann formalism finds a useful application in the *thermodynamic* context, as will be established below. In this case the metric geometry governed by $\mathscr{G}_\xi$ concerns thermodynamic responses, while the manifold ξ labels distinct states of thermodynamic equilibrium (each of which can exhibit its own local geometry of responses). The thermodynamic manifold ξ can be identified with Gibbs space, whose points lack an intrinsic metric significance but furnish the necessary parametrization of equilibrium states. The associated metric $\mathscr{G}_\xi$, on the other hand, will be derived from general criteria of equilibrium and stability (empirical laws of thermodynamics), which govern the possible responses of a thermodynamic system in any particular state of thermal equilibrium. A general thermodynamic theory for the totality of equilibrium states accessible to a given system would thus find its natural expression in the more general metric formalism of Riemann, but in the present article attention will essentially be confined to the local Euclidean geometry of an individual equilibrium state.

It is perhaps worth stressing that the notion of a metric space, as defined in Section III,A,1 and illustrated above, does *not* include the Minkowski space of four-dimensional space-time, or related non-Euclidean geometries. In the latter cases, one or more of the axiomatic requirements (3.1) is abandoned [particularly (3.1c)], and the resulting mathematical structure differs profoundly from that of a metric space, though of course a certain formal correspondence can be maintained.

B. Elements of Equilibrium Thermodynamics

1. *Macroscopic Thermodynamic State Properties*

Although thermodynamics unarguably concerns relationships among the macroscopic properties which describe a physical system in its various states of thermal equilibrium, it is a problem of somewhat greater subtlety to establish what precisely is meant by "macroscopic," "state," "equilibrium," etc., what constitutes a suitable set of "properties" for the description of such states, or how such properties are measured. As to the latter point, very little will be said here. While volume, pressure, mass, work, and related properties can be understood on a purely mechanical basis, such thermal properties as heat, temperature, and entropy are intimately bound up with thermodynamic theory itself, and indeed, much of the classical formalism of Clausius and Kelvin is concerned with establishing the operational significance of such properties and providing a theoretical basis for their practical measurement via thermometry and calorimetry. The zeroth law and the concepts of adiabatic and diathermal partitions can be employed to introduce such concepts in an orderly manner (cf. Landsberg, 1961), but for present purposes it is sufficient to recognize that practical experimental procedures exist for the determination of the various properties which will come under discussion.

The notion of a thermodynamic *state* and the associated set of macroscopic properties by which it is specified may be operationally understood as follows (cf. Landsberg, 1961, p. 2). Suppose that numerical values of a selected set of macroscopic properties of a thermodynamic system have been chosen, and that many copies of the system are now prepared having all the same values of these properties. If an independent observer can detect a macroscopic difference in any of these copies (other than spatial location, shape, etc.), the overall procedure can be repeated with the value of this new property being specified in each member of the ensemble. After a sufficient number of properties have thereby been added, it will eventually be found that no further macroscopic differences can be detected, and the copies are then said to share the same thermodynamic state. Empirically, it is found that only a small, fixed number of distinct macroscopic properties is required to specify any such state, and the number can in fact be anticipated from a simple rule involving the number of phases and chemical components. From a microscopic point of view, it is perhaps one of the most remarkable observations concerning thermodynamic systems that this number (of degrees of freedom) is quite small, typically two or three rather than, say, 10^{23}.

The notion of *macroscopic* systems and properties may be made more precise in a similar manner. Let the ensemble of copies be grouped into pairs, triples, or other higher multiples, and let the partitions that isolate each member of the multiple from the others be removed to produce new composite systems. The properties of the composite systems thus formed may be compared with their former values in the separated subsystems. It may thereby prove possible to distinguish two classes of properties: (1) those (intensive) whose values are, within specified limits, numerically the same as in the original units, and (2) those (extensive) whose values are, within the same limits, simple multiples of the original values by the number of subsystems in the composite, and which thus exhibit simple additive behavior. If the properties specifying the states of the original subsystems can all be chosen to be of the second class, the subsystems (or their multiples) may properly be described as macroscopic, as may the properties specifying their state, for it is observed empirically that the extensive behavior, once established to a certain accuracy, will persist in all larger composite systems.† Generally speaking, it is found that many physical systems whose dimensions are large compared to atomic dimensions satisfy such criteria and can thus be described by the formalism of equilibrium thermodynamics. Conversely, the extensive character of a set of macroscopic state properties is a fundamental empirical element to be incorporated in this formalism.

Analogous procedures will suggest themselves by which one could distinguish the proper state properties from such nonconserved quantities as heat or work, and for similar matters of definition. Following Gibbs, we therefore assume that it is meaningful to associate such state properties as internal energy (U), entropy (S), volume (V), temperature (T), and pressure (P) with each equilibrium state, and seek now to describe the relationships among such properties that are characteristic of a chosen state.

2. *Potentials and Conjugate Variables*

It was emphasized by Gibbs that the internal energy U or entropy S can play the special role of a *thermodynamic potential* ϕ, whose functional dependence on the remaining extensive variables X_i,

$$\phi = \phi(X_1, X_2, \ldots, X_{c+2}) \tag{3.26}$$

is a fundamental equation for describing the equilibrium properties. Thus, if N_i denotes the quantity (e.g., mole number) of the ith chemical component, Eq. (3.26) could take either of the forms

$$U = U(S, V, N_1, N_2, \ldots, N_c) \tag{3.27a}$$

$$S = S(U, V, N_1, N_2, \ldots, N_c) \tag{3.27b}$$

where c is the number of independent components. The formalisms resulting from (3.27a) and (3.27b) are referred to as the energy representation and entropy representation, respectively, and each suffices for a complete and self-sufficient development of thermodynamics. Where there is need to be specific, the energy representation (3.27a) will be chosen, but we adopt the more general notation of Eq. (3.26) in order to emphasize the structural *roles* of the various quantities, rather than their particular identities.

In particular, each X_i of (3.26) is to be regarded more generally as any

† Note that these larger composite systems are regarded as being in the *same* thermodynamic state. That is, because of the simple way in which their properties are related, one does not distinguish macroscopic systems which differ merely in overall scale.

linear combination of the standard arguments listed in (3.27a)

$$\begin{aligned} X_1 &= a_{1s}S + a_{1v}V + \cdots + a_{1c}N_c \\ &\vdots \\ X_{c+2} &= a_{c+2,s}S + a_{c+2,v}V + \cdots + a_{c+2,c}N_c \end{aligned} \tag{3.28}$$

where the coefficients are arbitrary so long as the overall transformation is nonsingular,

$$\det \begin{vmatrix} a_{1s} & & a_{1c} \\ \vdots & & \vdots \\ a_{c+2,s} & \cdots & a_{c+2,c} \end{vmatrix} \neq 0. \tag{3.29}$$

These X_i's therefore exhibit the same extensive (additive) behavior as did the standard properties S, V, N_i, and in view of (3.29) the transformation (3.28) can, if desired, be inverted at any time to give the standard form (3.27a).†

Each of the extensive variables X_i can be associated through the thermodynamic potential ϕ with its *conjugate* intensive variable R_i by the equation‡

$$R_i = \left(\frac{\partial\phi}{\partial X_i}\right)_{X_1 \cdots X_{i-1}X_{i+1} \cdots X_{c+2}} = \left(\frac{\partial\phi}{\partial X_i}\right)_{\mathbf{X}}. \tag{3.30}$$

It was Gibbs who recognized that this conjugate association of pairs (X_i, R_i) provides convenient expressions for the temperature and pressure

$$(V, -P): \qquad P = -(\partial U/\partial V)_{S, N_1 \cdots N_c} \tag{3.31a}$$

$$(S, T): \qquad T = (\partial U/\partial S)_{V, N_1 \cdots N_c} \tag{3.31b}$$

and, moreover, leads naturally to the notion of *chemical potentials* μ_j, respectively conjugate to the remaining arguments N_j,

$$(N_j, \mu_j): \mu_j = (\partial U/\partial N_j)_{S, V, N_1 \cdots N_{j-1}N_{j+1} \cdots N_c} \tag{3.31c}$$

which makes possible a full chemical thermodynamics.

One should recognize that the conjugacy relations (3.30) already reflect *empirical* information concerning the mathematical "smoothness" of the potential ϕ. Of course, there is no general reason to assume that an *arbitrary* mathematical function ϕ should necessarily possess any of the derivatives that define the conjugate intensities, yet experience shows that, in the ther-

† As S, V, and the N_i are measured in different units, one may feel an instinctive unease in "adding apples and hummingbirds," as seems to be done in Eqs. (3.28). Some reflection will confirm, however, that such transformations of variables have a perfectly good meaning, both mathematically and physically, in the present context.

‡ As Eq. (3.30) indicates, a subscript $\mathbf{X}$ on a partial derivative abbreviates the set of state properties held constant for the partial differentiation, i.e., all X_j's except X_i itself. Of course, a symbol $\partial\phi/\partial X_i$ has no definite meaning until all the independent variables held constant are identified.

modynamic context, these derivatives indeed exist, and the associated intensive properties vary in a continuous, well-behaved way with changes of state,

$$R_i = R_i(X_1, X_2, \ldots, X_{c+2}). \tag{3.32}$$

3. *Criteria of Equilibrium and Stability*

The intensities (3.32) are observed to be sufficiently well-behaved functions of the X_i's to permit differentials of each R_i to be defined and manipulated according to the familiar rules of the partial differential calculus. Indeed, the first law—the observation that internal energy U is a conserved function of state, and thus has an *exact* differential—is expressed mathematically (see, e.g., Margenau and Murphy, 1956, p. 8) by the set of cross-differentiation relations

$$\left(\frac{\partial^2 \phi}{\partial X_i\, \partial X_j}\right)_{\mathbf{X}} = \left(\frac{\partial^2 \phi}{\partial X_j\, \partial X_i}\right)_{\mathbf{X}}, \qquad i, j = 1, 2, \ldots, c+2, \tag{3.33}$$

or, equivalently, in the form

$$\left(\frac{\partial R_i}{\partial X_j}\right)_{\mathbf{X}} = \left(\frac{\partial R_j}{\partial X_i}\right)_{\mathbf{X}}, \qquad i, j = 1, 2, \ldots, c+2 \tag{3.34}$$

which directly involves such intensity differentials. Note that in a mathematical sense, Eqs. (3.34) can be regarded as the conditions for the *existence* of some potential function $\phi = \phi(\mathbf{X})$ satisfying Eqs. (3.30), so that one may be led to develop the formalism entirely in terms of X_i's and R_j's.

One must consider in addition the constraints furnished by the second law, which establishes the *extremum* property of the potential function ϕ, and thus gives to states of thermodynamic equilibrium their characteristic stability. Whereas earlier workers had expressed this law in terms of the temporal increase of entropy in natural processes, Gibbs recognized that its consequences could be seen as well in the analytic properties of his fundamental equation (3.26) for a particular state. For inasmuch as the entropy of an isolated material system "strives toward a maximum" in processes leading toward stable equilibrium, it is a natural inference that the entropy function, in the final state thus attained, must have the mathematical character of a maximum. The extremum character of the entropy function can correspondingly be taken as the analytic criterion for the attainment of stable equilibrium, and thus as a definition of those systems whose properties are to be encompassed by the formalism of equilibrium thermodynamics.

While the foregoing statements (which concern maximization of the entropy at constant energy) are couched in the entropy representation (3.27b), equivalent observations concerning the minimization of energy at constant S follow from the definiteness of the temperature derivative (3.31b),† and hence allow one to proceed with either form of the potential ϕ.

The observed extremum character of the thermodynamic potential ϕ with respect to arbitrary infinitesimal variations δX_i in its arguments may be expressed very generally through a convexity relation of the form (see, e.g., Fisher, 1967; Stanley, 1971)

$$\phi(X_i) \leq \tfrac{1}{2}[\phi(X_i + \delta X_i) + \phi(X_i - \delta X_i)], \qquad \text{all } X_i. \tag{3.35}$$

But under continuity conditions which the potential is known to satisfy, the minimum property of ϕ can be expressed as well (Hardy *et al.*, 1967) in terms of the usual second-derivative condition‡

$$\left(\frac{\partial^2 \phi}{\partial X_i^2}\right)_{\mathbf{X}} = \left(\frac{\partial R_i}{\partial X_i}\right)_{\mathbf{X}} \geq 0, \qquad \text{all } X_i, \tag{3.36}$$

which expresses the upward curvature of ϕ with respect to small displacements about the isolated equilibrium state. Note that conditions (3.35), (3.36) are required to hold with respect to arbitrary small changes δX_i, i.e., for each value of the index i, and for every possible way of choosing the X_i's. Note also that (3.35), (3.36) impose only a criterion of local convexity on the potential ϕ, so that the formalism applies as well to systems that are only *meta*stable (as most systems must ultimately be), when such systems fulfill the other standard requirements for its application. Note finally that the second form of the inequality in (3.36) permits one to proceed without direct reference to ϕ itself, as suggested also by Eq. (3.34).

4. *Independent Intensive Properties*

The basic intensities (3.32), whose differentials are constrained to satisfy the first and second laws in the form of Eqs. (3.34), (3.36), are interrelated by the *Gibbs phase rule* in a way which will now be made explicit. The phase rule establishes that the number r of independent intensities (number of

† See, e.g., Callen (1960, pp. 85–90). A metric structure similar to that discussed below could also be constructed for negative temperature states as discussed by Ramsey (1956, 1974). Temperature of indefinite sign would require one to determine which of the extremum principles (that of the energy or the entropy) is the more fundamental, but would not otherwise interfere with the construction of a thermodynamic metric geometry in one or the other of the representations (3.27).

‡ Additional consequences of (3.35) can be inferred concerning higher order derivatives if some of these second derivatives vanish, as at a critical point. Note that conditions such as (3.34) and (3.36) implicitly restrict the thermodynamic potential ϕ to be one of the two Gibbsian forms (3.27), rather than, e.g., their Legendre transforms (enthalpy, free energies, etc.). While such Legendre transforms have many of the properties of a potential, their second derivatives would be undefined for some thermodynamic states of interest, and their Jacobian matrices would not properly reflect the thermodynamic dimensionality in the manner to be discussed below (Section III,B,4).

degrees of freedom) in a system consisting of c independent chemical components and v distinct phases is

$$r = c - v + 2. \tag{3.37}$$

This rule—perhaps more than the so-called "third law of thermodynamics" (Nernst postulate)—deserves to be accorded the stature of a fundamental thermodynamic law, for it establishes empirically based constraints on thermodynamic responses which are no less fundamental than those associated with the first and second laws. Indeed (as noted in Section III,B,1), the circumstance by which assembles of some 10^{23} microparticles become describable in terms of only a few degrees of freedom establishes one of the most interesting and characteristic features of a thermodynamic formalism.

The phase rule summarizes the following empirical observations:

(a) That the simplest (one-component, one-phase) thermodynamic systems have exactly *two* degrees of freedom;

(b) That each "independent" chemical component contributes an additional degree of freedom to a one-phase system;

(c) That the potential of a single phase is *extensive*, i.e., is mathematically a homogeneous function (of degree one) when expressed in terms of $c + 2$ extensive state properties X_i;

(d) That the potential of a multiphase system is the sum of those for the individual phases.

As is well known, observation (c) [which includes the consequences of (a) and (b)] allows one to write down a Gibbs–Duhem equation for each distinct phase, while (d) permits these to be combined to give the final form (3.37) of the phase rule.

The "independence" of a set of R_i's, which all depend functionally on the set of X_j's as in (3.32), is to be understood in the usual mathematical sense, i.e., as implying the nonvanishing of a Jacobian determinant for the set of r variables,

$$\frac{\partial(R_1, R_2, \ldots, R_r)}{\partial(X_1, X_2, \ldots, X_r)} \neq 0. \tag{3.38a}$$

The phase rule guarantees that such a choice of R_i's is always possible, and moreover, that any larger set of intensities is necessarily dependent,

$$\frac{\partial(R_1, R_2, \ldots, R_r, R_{r+1})}{\partial(X_1, X_2, \ldots, X_r, X_{r+1})} = 0, \qquad \text{all } X_{r+1}. \tag{3.38b}$$

Such independence conditions may be more succinctly summarized in terms of the $(c + 2) \times (c + 2)$ Jacobian matrix $\mathsf{G}^{(c+2)}$, whose elements are

$$(\mathsf{G}^{(c+2)})_{ij} = (\partial R_i/\partial X_j)_{\mathbf{X}}, \qquad i, j = 1, 2, \ldots, c + 2. \tag{3.39}$$

Since vanishing subdeterminants of (3.39) as in (3.38b) reflect the number of zero eigenvalues of $\mathsf{G}^{(c+2)}$, independence conditions such as (3.38) are equivalently expressed in the form

$$\text{rank}(\mathsf{G}^{(c+2)}) = r = c - v + 2. \tag{3.40}$$

This may be considered to be a more complete and rigorous statement of Gibbs' phase rule, making explicit the independence of the intensities enumerated by (3.37).

Equation (3.40) implies that there exist exactly v linearly independent vectors $\boldsymbol{\eta}^{(\lambda)}$, $\lambda = 1, 2, \ldots, v$, which are null eigenvectors of the Jacobian matrix $\mathsf{G}^{(c+2)}$,

$$\mathsf{G}^{(c+2)}\boldsymbol{\eta}^{(\lambda)} = 0, \qquad \lambda = 1, 2, \ldots, v. \tag{3.41}$$

Of course, these vectors are not unique, since any linear combination of vectors satisfying Eq. (3.41) would also be a solution of this equation. However, the v-dimensional subspace spanned by these null eigenvectors *is* unique and depends only on the phase homogeneity (c) and additivity (d) observations which led to the phase rule (3.40). To complete the mathematical expression of observations (a)–(d), it is therefore sufficient to specify a set of the null eigenvectors $\boldsymbol{\eta}^{(\lambda)}$ which can guarantee the phase homogeneity and additivity.

Let us first recall how the familiar conditions for phase coexistence arise from the phase–additivity observation (d). The potential, extensities, and intensities of the λth phase will be denoted respectively by

$$\phi^{(\lambda)}, X_i^{(\lambda)}, R_i^{(\lambda)} = (\partial\phi^{(\lambda)}/\partial X_i^{(\lambda)})_{\mathbf{X}^{(\lambda)}} \tag{3.42}$$

where, in the absence of restrictions on the exchange of quantity X_i among the subsystems,

$$\phi(X_1 \cdots X_{c+2}) = \sum_{\lambda=1}^{v} \phi^{(\lambda)}(X_1^{(\lambda)} \cdots X_{c+2}^{(\lambda)}) \tag{3.43a}$$

$$X_i = \sum_{\lambda=1}^{v} X_i^{(\lambda)}, \qquad i = 1, 2, \ldots, c+2. \tag{3.43b}$$

Equation (3.43a) implies $d\phi = \sum_\lambda d\phi^{(\lambda)}$, or

$$\sum_{i=1}^{c+2} \left(\frac{\partial\phi}{\partial X_i}\right)_{\mathbf{X}} dX_i = \sum_{\lambda=1}^{v} \sum_{i=1}^{c+2} \left(\frac{\partial\phi^{(\lambda)}}{\partial X_i^{(\lambda)}}\right)_{\mathbf{X}^{(\lambda)}} dX_i^{(\lambda)} \tag{3.44a}$$

which may be rewritten in the form

$$\sum_{i=1}^{c+2} \sum_{\lambda=1}^{v} (R_i - R_i^{(\lambda)})\, dX_i^{(\lambda)} = 0 \tag{3.44b}$$

with the help of (3.43b) and the definitions of R_i, $R_i^{(\lambda)}$. As (3.44) must hold for every possible change $dX_i^{(\lambda)}$, one concludes

$$R_i^{(\lambda)} = R_i, \qquad \text{each } i = 1, 2, \ldots, c+2; \quad \lambda = 1, 2, \ldots, \nu \tag{3.45}$$

which is the usual condition of equality of the intensive properties (temperature, pressure, chemical potentials) in every phase.

The "homogeneity" of each phase λ (in both the physical and mathematical sense) has as a consequence that an increase of each $X_i^{(\lambda)}$ by an amount proportional to its numerical value in that phase,

$$dX_i^{(\lambda)} = s^{(\lambda)} X_i^{(\lambda)} \qquad (s^{(\lambda)} > 0), \qquad i = 1, 2, \ldots, c+2 \tag{3.46a}$$

merely *scales* the overall phase by a factor $s^{(\lambda)}$, but does not affect its intensive properties,

$$dR_i^{(\lambda)} = 0 \qquad \text{under scaling} \tag{3.46b}$$

and hence leaves the equilibrium conditions (3.45) undisturbed. Accordingly, in a composite system of ν phases for which

$$dR_i = \sum_{j=1}^{c+2} \left(\frac{\partial R_i}{\partial X_j}\right)_{\mathbf{X}} dX_j, \qquad i = 1, 2, \ldots, c+2 \tag{3.47}$$

a set of "changes" dX_j taken proportional to $X_j^{(\lambda)}$, the quantity of X_j in the λth phase,

$$dX_j = s^{(\lambda)} X_j^{(\lambda)}, \qquad j = 1, 2, \ldots, c+2 \tag{3.48a}$$

will serve merely to increase the total quantity of phase λ, but will not affect any intensive properties of the composite system,

$$dR_i = 0, \qquad \text{under scaling of } \lambda\text{th phase.} \tag{3.48b}$$

Equations (3.47), (3.48) lead to the result, for each $\lambda = 1, 2, \ldots, \nu$,

$$0 = \sum_{j=1}^{c+2} \left(\frac{\partial R_i}{\partial X_j}\right)_{\mathbf{X}} s^{(\lambda)} X_j^{(\lambda)}, \qquad i = 1, 2, \ldots, c+2 \tag{3.49}$$

which implies (since $s^{(\lambda)} \neq 0$)

$$0 = \sum_{j=1}^{c+2} (\mathsf{G}^{(c+2)})_{ij} X_j^{(\lambda)}, \qquad i = 1, 2, \ldots, c+2; \quad \lambda = 1, 2, \ldots, \nu. \tag{3.50}$$

Equations (3.50) are precisely of the form (3.41), with the null eigenvectors $\boldsymbol{\eta}^{(\lambda)}$ now identified to have the elements

$$(\boldsymbol{\eta}^{(\lambda)})_j = X_j^{(\lambda)} = \begin{matrix}\text{numerical value of extensive} \\ \text{variable } X_j \text{ in phase } \lambda,\end{matrix} \tag{3.51}$$

$$j = 1, 2, \ldots, c+2; \quad \lambda = 1, 2, \ldots, \nu.$$

Equations (3.41) and (3.51) [together with the definition (3.39)] therefore make mathematically precise the substance of Gibbs' phase rule and the associated phase homogeneity and additivity observations which underlie a thermodynamic formalism.

In view of these considerations, it is evident that Eqs. (3.32) do not yet reflect properly the mathematical dependence on a set of r independent state properties, as required by the phase rule. The arguments X_i of the potential ϕ may evidently be divided into two categories: (i) those (r in number) that can serve as independent variables to uniquely define the equilibrium state, and (ii) those remaining ($v = c + 2 - r$ in number) that serve as scale factors to specify effectively the size of each phase, and will take on fixed constant values in all thermodynamic derivatives. The latter category may be distinguished by the symbols $\hat{X}_\kappa$, $\kappa = 1, 2, \ldots, v$,

$$\underbrace{X_1, X_2, \ldots, X_r;}_{\text{"variables" } \mathbf{X}} \qquad \underbrace{\hat{X}_1 = X_{r+1}, \hat{X}_2 = X_{r+2}, \ldots, \hat{X}_v = X_{r+v}}_{\text{"scale factors" } \hat{\mathbf{X}}} \tag{3.52}$$

so that Eqs. (3.26) and (3.30) take the respective forms

$$\phi = \phi(X_1, X_2, \ldots, X_r, \hat{X}_1, \ldots, \hat{X}_v) = \phi(\mathbf{X}, \hat{\mathbf{X}}) \tag{3.53}$$

$$R_i = (\partial\phi/\partial X_i)_{\mathbf{X}, \hat{\mathbf{X}}}, \qquad i = 1, 2, \ldots, r \tag{3.54a}$$

$$R_{r+\kappa} = (\partial\phi/\partial \hat{X}_\kappa)_{\mathbf{X}, \hat{\mathbf{X}}}, \qquad \kappa = 1, 2, \ldots, v. \tag{3.54b}$$

Although the definitions (3.54) of the intensities require implicit specifications of the scale factors held constant, the numerical values of these intensities are independent of the values of the $\hat{X}_\kappa$'s, so that Eq. (3.32) is to be replaced by

$$R_i = R_i(X_1, X_2, \ldots, X_r) = R_i(\mathbf{X}), \qquad i = 1, 2, \ldots, r \tag{3.55a}$$

$$R_{r+\kappa} = R_{r+\kappa}(X_1, X_2, \ldots, X_r) = R_{r+\kappa}(\mathbf{X}), \qquad \kappa = 1, 2, \ldots, v \tag{3.55b}$$

to reflect properly their mathematical dependence on only the smaller number of independent variables. Of course, the distinction between variables and scale factors in (3.52) is always to be made in such a way that the conjugate intensities R_i of (3.55a) are properly independent in the sense of (3.38a).

5. *Linearity of Intensity Differentials*

Let ξ denote the set of numerical values of the X_i's that uniquely labels the particular equilibrium state under consideration. It may now be observed that the empirical laws summarized in Eqs. (3.34), (3.36), (3.40) have been expressed in a form that involves only the *first differentials* dR_i of intensive state properties, evaluated at the point ξ. This observation leads one to analyze the algebra of these differentials in somewhat greater detail, since it is only this differential aspect of the intensity functions that is subject to thermodynamic constraints and is therefore of direct concern in a thermodynamic formalism. This is fortunate, since the full mathematical form of the potential function (3.53) or the intensity functions (3.55) would seldom be available for real physical systems. Indeed, plausible mathematical assumptions on these functions concerning any but their lowest order derivatives may lead to sharp contradictions with experiment.† A first consequence of this observation is that two "different" intensities $R(X_1, X_2, \ldots, X_r) \neq \tilde{R}(X_1, X_2, \ldots, X_r)$ whose first differentials are equal in the state ξ,

$$dR(\xi) = d\tilde{R}(\xi) \tag{3.56}$$

are essentially equivalent for describing the thermodynamics of that state, and might be used interchangeably in calculating its properties.

The algebra satisfied by the intensity differentials dR_i is that determined by the "chain rules" and other familiar rules of the partial differential calculus. Let us consider some new set of properties $\tilde{R}_1, \tilde{R}_2, \ldots, \tilde{R}_r$ to be used in place of the reference set $R_1, R_2, \ldots, R_r$ for describing the state ξ. The $\tilde{R}_i$'s are required to satisfy the usual condition of independence

$$\left.\frac{\partial(\tilde{R}_1, \tilde{R}_2, \ldots, \tilde{R}_r)}{\partial(R_1, R_2, \ldots, R_r)}\right|_\xi \neq 0 \tag{3.57}$$

but are otherwise arbitrary. Condition (3.57) guarantees, by the *implicit function theorem* (Kaplan, 1952), the existence of explicit invertible functional relationships

$$\tilde{R}_i = \tilde{R}_i(R_1, R_2, \ldots, R_r) \tag{3.58}$$

in the neighborhood of the state ξ in question. In particular, their first differentials satisfy simple linear equations

$$d\tilde{R}_i = \sum_{j=1}^{r} a_{ij}\, dR_j, \qquad i = 1, 2, \ldots, r \tag{3.59}$$

where the coefficient a_{ij} has the numerical value

$$a_{ij} = \left.\frac{\partial \tilde{R}_i}{\partial R_j}\right|_\xi \tag{3.60}$$

and thus depends on the particular equilibrium state ξ as well as on the

† For example, the assumption of a low-order Taylor series expansion of the Gibbs or Helmholtz free energy in the neighborhood of a critical point (as in the Landau phenomenological theory of phase transitions) leads to disagreements with observed critical properties; see, e.g., Fisher (1967, pp. 659ff), Stanley (1971, Chapter 10).

particular choice of reference fields. These coefficients may conveniently be regarded as elements of a column vector $\mathbf{a}_i$, with $(\mathbf{a}_i)_j \equiv a_{ij}$. The thermodynamically significant differential properties of each $\tilde{R}$ may then be labeled by the vector $\mathbf{a}$, which allows them to be expressed in terms of the chosen reference set,

$$d\tilde{R} \equiv dR_{\mathbf{a}} = \sum_{i=1}^{r} (\mathbf{a})_i \, dR_i. \tag{3.61}$$

Such linear relationships may be summarized quite generally in the form

$$dR_{\lambda\mathbf{a}+\mu\mathbf{b}} = \lambda \, dR_{\mathbf{a}} + \mu \, dR_{\mathbf{b}} \tag{3.62}$$

for arbitrary (r-component) vectors $\mathbf{a}$, $\mathbf{b}$ and real scalars λ, μ. Note that the relationships (3.58) among intensities need not necessarily be linear, but those (3.59) or (3.62) among their first differentials are always rigorously so.

In view of (3.62), the intensity differentials evidently have the properties of a linear manifold. However, it is necessary to verify that each intensity $\tilde{R}$ thus defined is properly related to some conjugate extensive variable $\tilde{X}$, and that every choice of $\tilde{X}$ gives rise to an $\tilde{R}$ which belongs to the manifold. To see this, let the general coefficients a_{ij} in (3.60) be regarded as elements of a matrix A,

$$(\mathsf{A})_{ij} = a_{ij}. \tag{3.63}$$

We denote by $\bar{\mathsf{A}}$ the corresponding inverse transpose ("*conjugate*") matrix, with elements $\bar{a}_{ij}$,

$$(\bar{\mathsf{A}})_{ij} = (\mathsf{A}^{-t})_{ij} = (\mathsf{A}^{-1})_{ji} = \bar{a}_{ij} \tag{3.64}$$

and take the $\tilde{X}_i$'s to be defined as

$$\tilde{X}_i = \sum_{j=1}^{r} \bar{a}_{ij} X_j, \qquad i = 1, 2, \ldots, r. \tag{3.65}$$

It is then an easy exercise in the partial differential calculus to show that

$$\left(\frac{\partial \phi}{\partial \tilde{X}_i}\right)_{\tilde{\mathbf{X}}, \hat{\mathbf{X}}} = \sum_{j=1}^{r} a_{ij} R_j \tag{3.66}$$

and hence that $\partial\phi/\partial\tilde{X}_i$ "equals" $\tilde{R}_i$ for thermodynamic purposes, since their differentials coincide in the state $\boldsymbol{\xi}$. That these conjugate extensive variables $\tilde{X}_i$ exist is precisely guaranteed by the independence condition (3.57), which assures the existence of the conjugate matrix $\bar{\mathsf{A}}$.

C. Thermodynamics and Geometry

When the empirical laws of thermodynamics [as mathematically summarized in Eqs. (3.34), (3.36), (3.40), (3.62) of Section III,B] are compared with the axiomatic requirements of an abstract r-dimensional metric geometry [as summarized in Eqs. (3.1), (3.20) of Section III,A], a far-reaching structural similarity can be perceived (Weinhold, 1975a), as the notation has to some extent been chosen to anticipate. It was previously pointed out that Eq. (3.62) invests the intensity differentials dR_i with the properties of a linear manifold, so that one may associate with each intensive variable R_i an abstract "vector" symbol $\mathscr{R}_i$,

$$\mathscr{R}_i \Leftrightarrow dR_i. \tag{3.67a}$$

More importantly, one may now attribute to these vectors a *metric* character by defining a scalar product

$$\langle \mathscr{R}_i | \mathscr{R}_j \rangle = (\partial R_i / \partial X_j)_{\mathbf{X}} \tag{3.67b}$$

based on measurable values of thermodynamic response functions.

Unlike previous examples (3.22)–(3.24) of metric geometries, the thermodynamic vector geometry (3.67) does *not* manifestly possess the distributive, symmetric, and positive properties (3.1a)–(3.1c) which guarantee its Euclidean character. Indeed, one could imagine many general "potentials" ϕ and associated "intensities" R_i for which (3.67b) would be completely meaningless. However, it will now be shown that the additional constraints imposed on ϕ by the empirical laws of thermodynamics are precisely those which will insure that (3.67b) is meaningful in a fully Euclidean sense.

Thus, with the definition (3.67b) one can readily recognize that the first law (3.34) insures the symmetric property (3.1b) of the scalar product,

$$\left(\frac{\partial R_i}{\partial X_j}\right)_{\mathbf{X}} = \left(\frac{\partial R_j}{\partial X_i}\right)_{\mathbf{X}} \Rightarrow \langle \mathscr{R}_i | \mathscr{R}_j \rangle = \langle \mathscr{R}_j | \mathscr{R}_i \rangle, \tag{3.68}$$

while the second law (3.36) similarly insures the positive property (3.1c),

$$\left(\frac{\partial R_i}{\partial X_i}\right)_{\mathbf{X}} \geq 0 \Rightarrow \langle \mathscr{R}_i | \mathscr{R}_i \rangle \geq 0, \tag{3.69}$$

for every vector of the space. The differential properties of state functions summarized in (3.62) will always insure the distributive property (3.1a),

$$dR_{\lambda\mathbf{a}+\mu\mathbf{b}} = \lambda\, dR_{\mathbf{a}} + \mu\, dR_{\mathbf{b}} \Rightarrow \langle \mathscr{R}_i | \lambda \mathscr{R}_j + \mu \mathscr{R}_k \rangle = \lambda \langle \mathscr{R}_i | \mathscr{R}_j \rangle + \mu \langle \mathscr{R}_i | \mathscr{R}_k \rangle$$
$$\text{(e.g., with } \mathscr{R}_j = \mathscr{R}_{\mathbf{a}},\quad \mathscr{R}_k = \mathscr{R}_{\mathbf{b}}). \tag{3.70}$$

The dimensionality of this metric space is determined by noting that the independence conditions (3.38) or (3.40) are precisely the dimensional conditions (3.8), (3.9) on the metric determinant of the space $\mathfrak{M}_r$,

$$\frac{\partial(R_1, R_2, \ldots, R_s)}{\partial(X_1, X_2, \ldots, X_s)} = \begin{cases} 0 & \text{if } s > r \\ \text{nonzero} & \text{if } s = r \end{cases} \Rightarrow \mathscr{R}_i \in \mathfrak{M}_r. \tag{3.71}$$

Finally, if $\mathscr{R}_i$ is any *non*zero vector, the phase rule in the form (3.38a) [or (3.8)] guarantees that another $r - 1$ nonzero vectors can be found for which the associated metric matrix $\mathsf{G}^{(r)}$

$$(\mathsf{G}^{(r)})_{ij} = \langle \mathscr{R}_i | \mathscr{R}_j \rangle, \qquad i, j = 1, 2, \ldots, r \tag{3.72}$$

is nonsingular. But this implies, by the well-known theorem of Frobenius (Mirsky, 1955, p. 400), that no diagonal element of this matrix can be zero, and hence that

$$\langle \mathscr{R}_i | \mathscr{R}_i \rangle \neq 0 \quad \text{unless} \quad \mathscr{R}_i = 0, \tag{3.73}$$

as required to complete the positivity condition (3.1c). Thus, the r-dimensional Euclidean character of the thermodynamic geometry (3.67) is a necessary consequence of general thermodynamic principles.

With this assurance, the general formalism of metric spaces can be taken over intact from Section III,A to develop thermodynamic theory in a new geometrical representation. The natural mathematical tools in this representation are the algebra and geometry of Euclidean vectors rather than the partial differential calculus of potential functions. The thermodynamic derivatives (response functions) become geometrical scalar products, related to lengths and angles of Euclidean vectors; the thermodynamic Jacobians of Eq. (3.38) become the fundamental metric of Eq. (3.9); "laws of thermodynamics" become merely the "rules of geometry"; and, in general, thermodynamic identities and theorems become identities and theorems of Euclidean geometry. The translation from the geometrical to the analytical representation is always possible through Eq. (3.67), so that, in principle, results derived from one formalism could always be recognized in the other. Yet the ease of carrying out vector and matrix algebra in Euclidean vector spaces—as against corresponding manipulations of partial differential quantities—will recommend the geometrical formalism in many situations, such as those involving complex chemical systems of several components and phases. In such cases a thermodynamic analysis may be carried through (Weinhold, 1976c) with a completeness considerably beyond that practically achievable in the usual formalism.

It will be evident from Eqs. (3.68)–(3.73) that the r-dimensional metric properties of the thermodynamic vectors (3.67a) are also sufficient to recover the principal empirical laws governing relationships among thermodynamic derivatives, as enumerated in Section III,B. Of course, a specific structure—that specified by Eq. (3.51)—is required of the null eigenvectors $\boldsymbol{\eta}^{(\lambda)}$ of Eq. (3.21) in order to properly insure the homogeneity and additivity of distinct phases. Moreover, it remains to be clarified by what rule of vector addition (or otherwise) the thermodynamic vectors of separated phases are related to those of a final composite system. Nevertheless, it is clear that a geometrical representation has little mathematical structure that is specific to thermodynamics, and that one gains a considerable formal simplification by exploiting the interesting relationship which unites the first and second laws with other spoken and unspoken empirical observations into a single mathematical framework of surprising simplicity and ubiquity—elementary Euclidean geometry.

Consider, for example, the behavior of a pure fluid or solid near absolute zero. The temperature vector $\mathscr{T} = \mathscr{R}_1$ (with pressure $\mathscr{P} = -\mathscr{R}_2$) is in this case related to the constant-volume heat capacity C_V by (Weinhold, 1975c)

$$\langle \mathscr{R}_1 | \mathscr{R}_1 \rangle = T/C_V. \tag{3.74}$$

By Eq. (3.73), the vanishing of this ratio would imply the vanishing of the temperature vector itself and hence a violation of Gibbs' phase rule (since the metric dimensionality would thereby be reduced from two to one). The heat capacity C_V must therefore approach zero as $T \to 0$ in such a way that the ratio (3.74) remains finite,

$$C_V \to 0 \qquad \text{as} \quad T \to 0. \tag{3.75}$$

Similar considerations may be used to adduce the vanishing of C_p and the heat of pressure variation, Γ_V, near absolute zero. Equation (3.75) shows, for example, that the classical Dulong–Petit law (Glasstone, 1946) of specific heats must give a violation of the laws of thermodynamics at sufficiently low temperatures. While conclusions such as (3.75) are usually based on a so-called "third law of thermodynamics," they are derived here without previous reference to such a law, and are thus implicit consequences of broader thermodynamic principles when the metric character of the latter is taken fully into account.

D. Geometrical Representation of Thermodynamics

Various features of the geometrical representation of equilibrium thermodynamics have been outlined and illustrated in a series of papers (Weinhold, 1975b-d, 1976c), which may be briefly summarized.

The Gibbs–Duhem equation and related questions of phase homogeneity were geometrically analyzed in Paper II (Weinhold, 1975b). It was shown that the Gibbs–Duhem equation arises from the linear dependence of $r + 1$ vectors in an r-dimensional space and that the major features of the geometrical representation might be applied intact to systems where the usual homogeneity assumptions are significantly weakened. Indeed, the successive linear-dependence conditions arising with successive numbers of phases could *each* be associated with a form of generalized homogeneity of the thermodynamic potential, only one of these conditions coinciding with the ordinary Gibbs–Duhem equation (when the phases are homogeneous in the usual sense).

In Paper III (Weinhold, 1975c) the formal structure of the geometrical formalism was described, with particular emphasis being placed on the metric matrix G, whose $r(r+1)/2$ independent elements completely specify the thermodynamic geometry. While intensity variables R_i are associated with basis vectors $\mathscr{R}_i$ of the abstract metric space $\mathfrak{M}_r$ as in (3.67), the conjugate extensive variables X_i were shown to be similarly associated with the biorthogonal set of vectors $\mathscr{X}_i$ satisfying

$$\langle \mathscr{R}_i | \mathscr{X}_j \rangle = \delta_{ij}, \qquad i, j = 1, 2, \ldots, r. \tag{3.76}$$

Away from metric singularities, the formalism is found to exhibit a far-reaching symmetry between the conjugate $\mathscr{R}_i$'s and $\mathscr{X}_j$'s, and the general transformation theory of $\mathfrak{M}_r$ similarly exhibits this symmetry when the notion of a "conjugate" (inverse transpose) matrix is introduced. Moreover, a set of self-conjugate ("normal") variables can be defined (and related to the eigenvectors of G) for which the thermodynamic response functions are completely uncoupled, and the thermodynamic analysis reduces to a particularly simple form. The geometrical content of various thermodynamic theorems and identities is exhibited, particularly the thermodynamic stability conditions, which arise as instances of the geometric requirement $\cos^2 \theta = 1$ for various thermodynamic angles θ.

In Paper IV (Weinhold, 1975d), matrix–algebraic techniques were employed to give new methods for the evaluation of general thermodynamic derivatives. These improve upon previously available techniques (Bridgman, 1925; Shaw, 1935) for the special case $r = 2$, but unlike these previous techniques, can be extended easily to systems of greater chemical complexity. Such procedures take advantage of the fact that increased chemical complexity affects the dimensionality r (and hence the dimensions of the various vectors and matrices), but does not affect the fundamental mathematical form of the general geometrical relationships.

In Paper V (Weinhold, 1976c), special properties of heterogeneous systems were described geometrically. Each new phase is associated with a loss of thermodynamic dimensionality and a corresponding null eigenvector of the thermodynamic metric matrix, which may be discussed in terms of invariants ("symmetries") of the system—sets of changes in the extensive variables X_i which leave the thermodynamic state unaltered. The vector equations furnished by these invariants lead to a set of coexistence coefficients—simple determinantal functions of the entropies, volumes, and extensive properties of the coexisting phases. From these coefficients, general expressions for the slopes of coexistence boundaries in phase diagrams of complex systems (generalized Clapeyron equations) and general conditions for the stationary points in such diagrams (generalized Gibbs–Konowalow laws) can be derived, both in a form applicable to systems of arbitrary complexity.

IV. Related Topics

A. Displacement Variables of Baur, Jordan, Jordan, and Mayer

In 1965, Baur, Jordan, Jordan, and Mayer (hereafter, BJJM) published a paper entitled "Towards a Theory of Linear Nonequilibrium Statistical Mechanics" in which the vectorial character of the thermodynamic formalism was suggested from a statistical–mechanical origin. Because the BJJM paper seems not to have been widely recognized among thermodynamicists,† and because its purely thermodynamic results bear strongly on our own, we review its content in some detail.

In the introduction, the authors describe their objective as "attempting to construct a general method for treating the behavior of systems approaching equilibrium in the realm of small displacements from equilibrium; the realm that is generally described as linear. The aim was to develop a method of comparable generality and rigor to that of Gibbs in the treatment of equilibrium systems." The paper consists of nineteen sections, the first eight of which (summarized in Section IX) concern static equilibrium and fluctuations from the viewpoint of classical statistical mechanics. Although the nominal viewpoint is microscopic and statistical mechanical, certain results in their Section III are of a purely thermodynamic character, dependent only on empirical macroscopic properties of equilibrium systems.

The BJJM paper begins, after an introductory section, with a review of the Gibbs formalism in the entropy representation, taken in a dimensionless form,

$$\sigma = -S/k = \sigma(U, V, N_1, N_2, \ldots) = \sigma(\{X_i\}) \tag{4.1}$$

where k = Boltzmann constant. A statistical mechanical expression is then given for each extensive variable X_i (except V) as an integral over the ensemble probability distribution W of a suitable function $\chi_i = \chi_i(p^{(N)}, q^{(N)})$ of the phase-space coordinates for an N-particle system, viz.,

$$X_i = \mathscr{I} W \chi_i, \tag{4.2}$$

where $\mathscr{I}$ is an integration operator which sums over particle numbers and integrates over all coordinates in phase space.

† Science Citation Index since 1967 lists three references to this work (by papers related to light scattering). The BJJM paper was kindly brought to the author's attention by Professor M. Baur.

After introducing Legendre transformations of the characteristic function σ, BJJM note that the minimum property of σ (or of its Legendre transforms) at equilibrium implies that

$$\sigma_{ii} \equiv N(\partial^2\sigma/\partial X_i^2) \geq 0. \tag{4.3}$$

It is then introduced as "a corollary" of Eq. (4.3) that

$$\sigma_{ij} \equiv N(\partial^2\sigma/\partial X_i\, \partial X_j) < (\sigma_{ii}\sigma_{jj})^{1/2} \tag{4.4}$$

and it is concluded that "[t]he relations of Eqs. [(4.3)] and [(4.4)] are those of scalar products of vectors." This leads them to write equations of the form

$$\mathbf{i}_\alpha \cdot \mathbf{i}_\beta = \sigma_{\alpha\beta} \tag{4.5}$$

where the $\mathbf{i}_\alpha$ are described as "basis vectors in a space of as many dimensions as there are values of α."

Since the validity of the "corollary" (4.4) is not established, nor are other necessary properties of a scalar product (especially, its distributive property) mentioned, it may be surmised that Eq. (4.5) was more directly motivated by statistical-mechanical expressions such as that given ten pages later (p. 113),

$$\langle F \cdot G \rangle = N_0^{-1} \mathscr{I} W_{\text{eq}} F \cdot G. \tag{4.6}$$

Since $\mathscr{I}$ is a linear operator and W_{eq} is nonnegative, the expression on the right can readily be shown to have all necessary properties of a scalar product, provided that W_{eq} and phase-space functions F, G appropriate to each thermodynamic variable can be defined, and the resulting integral exists. But whatever its origin, Eq. (4.5) is essentially the fundamental scalar product that was subsequently (and independently) identified by Weinhold from purely thermodynamic considerations.

An interpretation of the vectors defined through Eq. (4.5) is given by BJJM in terms of fluctuating displacements from the equilibrium state, following a treatment of Callen (1960). Displacements of the extensive variables from equilibrium are represented as coefficients of the vectors $\mathbf{i}_\alpha$, while those of the intensive variables are coefficients of the "reciprocal" (conjugate) set of vectors. Attention is thereby drawn to the "orthonormal" (normal) variables, in terms of which these fluctuations become independent, and the phenomenological Onsager flux-gradient matrix of their dynamical evolution can be diagonalized. Small displacements of these independent orthonormal variables from equilibrium therefore decay independently with a single exponential factor $\exp(-t/\tau)$.

In the BJJM treatment it is the independent displacement variables that come to the fore, equations such as (4.5) being employed primarily to establish their desirable properties for treating the spontaneous decay of nonequilibrium displacements by statistical mechanics. The emphasis throughout is on statistical mechanics rather than thermodynamics.† Given this emphasis, it is perhaps not surprising that the general thermodynamic significance of this analogy—its origin, scope, and consequences—was not pursued. However, the BJJM paper clearly anticipates important aspects of the metric geometry of equilibrium thermodynamics and serves to complement the later work by suggesting how the geometrical picture is to be extended into the near-equilibrium domain of irreversible thermodynamics and how the macroscopic metric geometry might be traced to a statistical-mechanical origin. The BJJM treatment also suggests (J. E. Mayer, private communication) how a closer analogy might be established with the formalism of quantum mechanics, based on orthonormal functions in phase space (rather than Hilbert space) giving rise to thermodynamic average values.

B. Geometry in Tisza's Formalism

Mention may also be made of certain questions raised by a thermodynamic formalism (called MTE theory—Macroscopic Thermodynamics of Equilibrium) as developed by Tisza (1951, 1961, 1966). Tisza's objective was to put Gibbs' theory into an axiomatic form, as Carathéordory had done for the formalism of Clausius and Kelvin. The resulting formal structure therefore has considerable overlap with the Gibbsian formalism and has sometimes been termed "neo-Gibbsian." In these works, Tisza makes reference to geometrical aspects of the Gibbsian formalism in a way that conflicts somewhat with the analysis of Section III. He calls particular attention to the so-called "stiffness matrix," which was used (as by Gibbs himself) to analyze questions of stability, and was referred to as the "metric matrix" in Section III,A,2. Tisza considers the reduction of this matrix to diagonal form to deduce stability conditions, employing for this purpose a special class of unimodular transformations whose principal minors are all unity—a so-called "restricted equiaffine group." The general significance of such transformation matrices may be questioned, for the reduction of the stiffness quadratic form in this manner is not invariant to a relabeling of variables.

In the metric geometry of Section III, the corresponding discussion of stability (Weinhold, 1975c) leads naturally to the unitary transformation of the metric matrix to its diagonal matrix of eigenvalues, which are properly invariant. The associated eigenvectors are essentially the normal variables (Section III,D) or the independent orthonormal variables of Baur *et al.* (Section IV,A), whereas in Tisza's work the significance of the eigenvalue problem for G was specifically denied. The fact that the ordinary Gibbs

† For example, the authors write (p. 113): "Were our interest limited to equilibrium thermodynamics the analogy between the intensive and extensive variables and the two reciprocal basis vector sets might be considered good clean mathematical fun, but of little scientific value."

space lacks an intrinsic metric structure was properly noted, but the further assumption was made that no satisfactory metric could be defined on the manifold of states parametrized by a Gibbs space.† As we have seen in Section III, this stronger conclusion was not warranted.

C. Euler Topological Formula and Gibbs' Phase Rule

Gibbs' phase rule bears a tantalizing resemblance to an older theorem from the topological theory of three-dimensional polyhedra, which suggests the possibility of a quite different relationship between thermodynamics and geometry. This theorem, discovered in 1752 by Euler,‡ relates the number of vertices v, edges e, and faces f of any regular or irregular polyhedron by the formula

$$f = e - v + 2 \tag{4.7}$$

in which form the general similarity to the Gibbs phase rule

$$r = c - \nu + 2 \tag{4.8}$$

will be manifest. This similarity has previously provoked speculation and discussion by several authors (Rüdel, 1929; Levin, 1946; Klochko, 1949; Mindel, 1962; Rouvray, 1974).

Euler's formula (4.7) relates only to a singly connected polyhedron, i.e., one that could be continuously deformed, without tearing or pasting, onto the surface of a sphere. By contrast, a multiply connected polyhedron is topologically equivalent to (continuously deformable onto) a torus (doughnut) of one or more holes and would be described by a generalized Euler formula

$$f = e - v + 2 - 2h, \tag{4.9}$$

where h represents the number of holes ("genus") of the surface, and the faces f are all singly connected. Note that only the topological, but not the metric, characteristics of the polyhedra are of significance in applying such formulas; thus, the edges need not generally be straight lines, nor must the faces be planar.

An elegant generalization of Euler's formula for singly connected polyhedra in spaces of n dimensions was obtained by Schlaefli in 1852. The

† Thus, he states (Tisza, 1966, p. 106): "Neither the elementary, nor the Riemannian theory of curvature can be applied in Gibbs space, in which no physically meaningful metric is definable," and on p. 236 (cf. also related statements on pages 105, 241): "[The spaces of mathematical physics] are usually characterized by a metric based on length and/or orthogonality of the basis vectors. Neither of these concepts is definable in Gibbs space."

‡ The same relationship was noted by Descartes in 1640. For an account of the historical development of Euler's formula, see Courant and Robbins (1943) and Loeb (1976).

Euler–Schlaefli relation may be written in the general form (Loeb, 1976)

$$\sum_{i=0}^{n}(-1)^i n_i = 1 + (-1)^n \tag{4.10}$$

where n_i denotes the number of elements of dimension i on the polyhedron. Since vertices, edges, and faces are elements of dimensionality 0, 1, and 2, respectively, and since the closed, singly connected surface of a polyhedron divides three-dimensional space into exactly two cells (inside and outside), one has, for the three-dimensional polyhedra under discussion,

$$n_0 = v, \qquad n_1 = e, \qquad n_2 = f, \qquad n_3 = \text{No. of cells} = 2, \tag{4.11}$$

so that Eq. (4.10) reduces in the case $n = 3$ to the special form (4.7) noted previously.

A simple proof of the Euler topological formula (4.7) may be given by induction. Note first that it represents no loss of generality to consider all faces of the polyhedron to be triangular, for if any face has more than three edges and vertices, one could imagine the corresponding polyhedron having an additional vertex on or above this face and connected to each of its (say) g vertices by a new edge so as to replace the original face by g new triangular faces; in this modified polyhedron one would have

$$v' = v + 1 \tag{4.12a}$$

$$e' = e + g \tag{4.12b}$$

$$f' = f + g - 1 \tag{4.12c}$$

so that $v' - e' + f' = v - e + f$ is left unchanged in replacing the original faces by triangular faces. Consider then any such polyhedron of v vertices for which Euler's formula is known to be correct, e.g., a tetrahedron, with $v = 4$. The effect of an additional vertex is to add three new edges and to replace the original face by three new faces, thereby to give

$$v' = v + 1 \tag{4.13a}$$

$$e' = e + 3 \tag{4.13b}$$

$$f' = f + 2. \tag{4.13c}$$

The relationship $v' - e' + f' = v - e + f$ then establishes that Euler's formula holds also for the polyhedron of $v + 1$ vertices, and hence, by induction, for any larger number of vertices.

If one attempts to pattern a proof of the Gibbs phase rule along the lines of the foregoing proof of Euler's formula, some basic differences can be noted. If, in the usual manner, one proceeds inductively on the number of phases ν, then each new phase is to be associated with an additional Gibbs–

Duhem (homogeneity) relationship among the $c + 2$ intensive variables which reduces the number of degrees of freedom by one, so that

$$v' = v + 1 \tag{4.14a}$$

$$c' = c \tag{4.14b}$$

$$r' = r - 1. \tag{4.14c}$$

From (4.13) and (4.14) it is evident that the inductive sequence has a quite different pattern in the two cases, and there is apparently no simple correspondence between elements of Eqs. (4.7) and (4.8) that would bring these two patterns into coincidence. For example, there are no allowable combinations of (v, e, f) for which either v or e or f is less than four [the lowest possible polyhedra correspond to $(v, e, f) = (4, 6, 4), (5, 8, 5), (5, 9, 6)$, etc.], but there are many such combinations of (v, c, r), including eight allowable combinations in which *all three* values of v, c, and r are less than four.

Mindel (1962) has argued, however, that Euler's formula and Gibbs' phase rule could be related by a correspondence of the form:

Topological interpretation		Thermodynamic interpretation	
faces f	$=$	$c(v - 1)$	(4.15a)
edges e	$=$	$v(c - 1)$	(4.15b)
vertices v	$=$	r ("all possible variations of the system").	(4.15c)

The Euler topological constraint $v - e + f - 2 = 0$ then becomes, in the thermodynamic interpretation,

$$r - v(c - 1) + c(v - 1) - 2 = r - c + v - 2 = 0 \tag{4.16}$$

which is the usual phase rule.

However, it can be recognized that the association required by Eqs. (4.15) cannot generally be acceptable either mathematically or physically. For example, a thermodynamic system of one phase or one component would seem by (4.15a) and (4.15b) to correspond to polyhedra with *no* faces or edges, which is mathematically meaningless. Furthermore, many polyhedra that are mathematically acceptable could not correspond to any realizable thermodynamic system. Thus, if c is eliminated from Eqs. (4.15a) and (4.15b), one derives an equation for the supposed number of phases corresponding to given numbers of edges and faces, viz.,

$$v = \tfrac{1}{2}\{f - e + 1 + [(f - e + 1)^2 + 4e]^{1/2}\}. \tag{4.17}$$

If this formula is applied to an octahedron ($v = 6, e = 12, f = 8$), the number of phases would appear to be *irrational*, a physically meaningless result. A similar situation arises for many other elementary polyhedra. Thus, the reasoning leading to Eqs. (4.15) cannot be correct.

In summary, there seems no present evidence to support the conjecture of a deeper relationship between Euler's formula and the Gibbs phase rule, and considerable reason to suppose their resemblance to be superficial and coincidental. Even this resemblance disappears if one considers the more general Euler–Schlaefli relationship (4.10), of which Euler's theorem is but the three-dimensional special case, or if one considers surfaces of multiple connectivity. Thus, no hidden topological or geometrical significance (beyond that described in previous sections) need be attributed to Gibbs' phase rule on the basis of this resemblance.

D. Other Geometrical Aspects of Thermodynamics

Mnemonic devices for representing Maxwell identities and other thermodynamic relationships have often been constructed from squares, octahedra, and other simple geometrical figures. A familiar example is the thermodynamic square illustrated below,

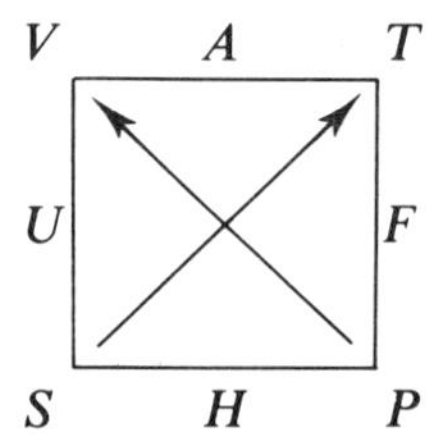

$$H = U + PV$$
$$A = U - TS$$
$$F = H - TS$$

whose use is described, e.g., by Callen (1960, pp. 119ff.). Such figures are sometimes called Born diagrams (see, e.g., Bowley, 1969; Tisza, 1966, p. 64) because they are said (Prins, 1948; Tisza, 1966) to have been used by Max Born in his thermodynamic lectures at Göttingen as early as 1929. However, the first published account of these squares and their underlying theory was given by Koenig (1935), who subsequently developed an analogous thermodynamic octahedron (Koenig, 1937, 1972) for treating three pairs of conjugate variables.

Koenig's original (1935) work was based on a substitution group of order eight which served to group thermodynamic equations into families and which could be derived from a set of geometrical operations on a square. This in turn stimulated work by McKay (1935) on the inclusion of additional conjugate pairs, by Buckley (1944) on the derivation of the Koenig substitution group from Lie's theory of contact transformations, and by Hayes (1946) on an enlarged substitution group. Prins (1947, 1948) and

Callen (1960) subsequently gave further extensions of the mnemonics based on squares. Recently, Fox (1976) gave a more elaborate thermodynamic mnemonic based on the cuboctahedron, one of the thirteen regular Archimedean solids. However, like the previous mnemonics, it is essentially restricted to systems having only one component or phase.

Mention may also be made of certain geometrical rules concerning the possible dispositions of first-order coexistence lines in general phase diagrams. When the three coexistence lines emanating from a thermodynamic triple point are examined, it is found that the angle subtended by a pure phase cannot exceed 180°. This "180° rule" has been discussed recently by Wheeler (1974), who gives references to the extensive literature of the subject and offers a sharpened thermodynamic proof for various choices of independent variables.

In his discussion, Wheeler emphasizes the role of two "explicit auxiliary assumptions," viz., that the extensive state properties of distinct phases cannot all be identical and that the values of these properties for any pure phase do not depend on the path of approach to the triple point. Model thermodynamic potentials having "all of the requirements of thermodynamic stability" (i.e., convexity) but violating these auxiliary assumptions (and hence leading to violations of the 180° rule) are presented. However, the notion of a state property which depends on a path of approach, or of distinct phases which cannot be distinguished by at least one extensive state property is in conflict with general operational definitions of such state properties, as described in Section III,B,1. While these auxiliary assumptions are described as being satisfied "at almost all presently known triple points," it seems that they are rather essential prerequisites for a thermodynamic description, and that the 180° rule should be regarded as a rigorous consequence of general thermodynamic principles. That model thermodynamic potentials can be constructed that violate the 180° rule should emphasize that these model potentials, being subject to an incomplete set of thermodynamic constraints, may have little to do with the thermodynamic behavior of real physical systems.

Reference may finally be made to certain other papers concerning the role of geometrical methods in thermodynamics. A particularly elegant application of geometrical methods was made by Griffiths and Wheeler (1970) in their general analysis of critical behavior in multicomponent systems. A vector geometry of irreversible thermodynamics has been employed by Leff and Jones (1975), while Andresen *et al.* (1977) have recently investigated the efficiency of finite-time thermodynamic processes, making use of a special metric for the distances between thermodynamic states in process diagrams.

Scalar-product formulations relating to thermodynamic theory have been developed from statistical–mechanical considerations, as in the work of Wilcox (1968) and references therein. Reference may also be made to an early article of Kurnakov (1928), who discusses the role of geometry in thermodynamics and bonding theory in a more philosophical vein.

V. Concluding Remarks

The interplay of thermodynamic theory with pure geometry has often played an important role in advancing the mathematical description of thermal phenomena. The present article has described a sense in which equilibrium thermodynamic theory is related in a still more fundamental way than was earlier supposed to the structure of a pure metric (Euclidean) geometry. The intrinsic metric structure, which distinguishes the present work from earlier applications of "geometry" in thermodynamic theory, is generated by the thermodynamic laws themselves, which can be recognized to be in precise correspondence with the axiomatic requirements of a metric space. This structural identity suggests that thermodynamic theory may usefully be represented in a mathematical framework more akin to that of Euclidean geometry than to the partial differential calculus of potential functions which is usually employed. Such a geometrical representation has particular advantages in treating complex chemical systems of many components and phases, since the increase in complexity affects the *dimensionality* of the vector space, but does not otherwise alter the *form* of the fundamental metric relationships which connect the thermodynamic vectors. The geometrical representation thus complements the usual analytic representation and seems likely to play a useful role in the further evolution of thermodynamic theory.

ACKNOWLEDGMENTS

This research was made possible through the financial support of a Teacher-Scholar Grant from the Camille and Henry Dreyfus Foundation. General support from the staff and facilities of the Theoretical Chemistry Institute, supported by the National Science Foundation, is also gratefully acknowledged.

References

Andresen, B., Salamon, P., and Berry, R. S. (1977). *J. Chem. Phys.* **66**, 1571.

Baur, M., Jordan, J. R., Jordan, P. D., and Mayer, J. E. (1965). *Ann. Phys.* (*N.Y.*) **35**, 96–163.

Bowley, W. W. (1969). *Am. J. Phys.* **37**, 1066–1067.

Bridgman, P. W. (1925). "A Condensed Collection of Thermodynamic Formulas." Harvard Univ. Press, Cambridge, Massachusetts.

Buckley, F. (1944). *J. Res. Natl. Bur. Stand.* **33**, 213–233.

Callen, H. B. (1960). "Thermodynamics." Wiley, New York.

Courant, R., and Robbins, H. (1943). "What is Mathematics?" Oxford Univ. Press, London and New York.

Dennery, P., and Krzywicki, A. (1967). "Mathematics for Physicists," pp. 109ff. Harper, New York.

Dicke, R. H., and Wittke, J. P. (1960). "Introduction to Quantum Mechanics," p. 130. Addison-Wesley, Reading, Massachusetts.

Donnan, F. G., and Haas, A. (1936). "A Commentary on the Scientific Writings of J. Willard Gibbs." Yale Univ. Press, New Haven, Connecticut.

Fisher, M. E. (1967). *Rep. Prog. Phys.* **30**, 615–730.

Fox, R. F. (1976). *J. Chem. Educ.* **53**, 441–442.

Gibbs, J. W. (1873a). *Trans. Conn. Acad. Arts Sci.* **2**, 309–342.

Gibbs, J. W. (1873b). *Trans. Conn. Acad. Arts Sci.* **2**, 382–404.

Gibbs, J. W. (1876–1878). *Trans. Conn. Acad. Arts Sci.* **3**, 108–243 and 343–524.

Gibbs, J. W. (1928). "Collected Works." Longmans, Green, New York.

Glasstone, S. (1946). "Physical Chemistry," 2nd ed., pp. 413–416. Van Nostrand-Reinhold, Princeton, New Jersey.

Griffiths, R. B., and Wheeler, J. C. (1970). *Phys. Rev. A* **2**, 1047–1064.

Hardy, G. H., Littlewood, J. E., and Polya, G. (1967). "Inequalities." Cambridge Univ. Press, London and New York.

Hayes, W. D. (1946). *Q. Appl. Math.* **4**, 227–232.

Kaplan, W. (1952). "Advanced Calculus," pp. 90ff. Addison-Wesley, Reading, Massachusetts.

Klochko, M. A. (1949). *Izv. Sekt. Fiz.-Khim. Anal., Inst. Obshch. Neorg. Khim., Akad. Nauk SSSR* **19**, 82.

Koenig, F. O. (1935). *J. Chem. Phys.* **3**, 29–35.

Koenig, F. O. (1937). *J. Phys. Chem.* **41**, 597–620.

Koenig, F. O. (1972). *J. Chem. Phys.* **56**, 4556–4562.

Kurnakov, N. W. (1928). *Z. Anorg. Allg. Chem.* **169**, 113–139.

Landsberg, P. T. (1961). "Thermodynamics, with Quantum Statistical Illustrations." Wiley (Interscience), New York.

Leff, H. S., and Jones, G. L. (1975). *Am. J. Phys.* **43**, 973–980.

Levin, I. (1946). *J. Chem. Educ.* **23**, 183–185.

Loeb, A. L. (1976). "Space Structures, Their Harmony and Counterpoint," Chapter 3. Addison-Wesley, Reading, Massachusetts.

McKay, H. A. C. (1935). *J. Chem. Phys.* **3**, 715–719.

Margenau, H., and Murphy, G. M. (1956). "The Mathematics of Physics and Chemistry," 2nd ed., pp. 8–12. Van Nostrand-Reinhold, Princeton, New Jersey.

Maxwell, J. C. (1875). "Theory of Heat," 4th ed. Longmans, Green, New York.

Mendoza, E. (1960). *In* "Reflections on the Motive Power of Fire, and Other Papers on the Second Law of Thermodynamics" (E. Mendoza, ed.), p. xi. Dover, New York.

Mindel, J. (1962). *J. Chem. Educ.* **39**, 512–514.

Mirsky, L. (1955). "An Introduction to Linear Algebra." Oxford Univ. Press, London and New York.

Misner, C. W., Thorne, K. S., and Wheeler, J. A. (1975). "Gravitation." Freeman, San Francisco, California.

Mountain, R. D. (1974). *J. Wash. Acad. Sci.* **64**, 195–198.

Prins, J. A. (1947). *Physica (Utrecht)* **13**, 417–421.

Prins, J. A. (1948). *J. Chem. Phys.* **16**, 65–66.

Ramsey, N. F. (1956). *Phys. Rev.* **103**, 20–28.

Ramsey, N. F. (1974). *In* "Modern Developments in Thermodynamics" (B. Gal-Or, ed.), pp. 207–219. Wiley, New York.

Rouvray, D. H. (1974). *Chem. Br.* **10**, 11–18.

Rüdel, O. (1929). *Z. Elektrochem.* **35**, 54.

Russell, B. (1956). "Essay on the Foundations of Geometry," pp. 22–26. Dover, New York.

Shaw, A. N. (1935). *Philos. Trans. R. Soc. London, Ser. A* **234**, 299–328.

Shilov, G. E. (1961). "An Introduction to the Theory of Linear Spaces," pp. 135ff. Prentice-Hall, Englewood Cliffs, New Jersey.

Stanley, H. E. (1971). "Introduction to Phase Transitions and Critical Phenomena," pp. 28ff. Oxford Univ. Press, London and New York.

Thomson, J. (1871). *Proc. R. Soc. London* **20**, 1–8.

Thomson, J. (1872). *Philos. Mag.* [7] **43**, 227–234.

Tisza, L. (1951). *In* "Phase Transformations in Solids" (R. Smoluchowski, J. E. Mayer, and W. A. Weyl, eds.), pp. 1–37. Wiley, New York.

Tisza, L. (1961). *Ann. Phys.* (*N.Y.*) **13**, 1–92.

Tisza, L. (1966). "Generalized Thermodynamics." MIT Press, Cambridge, Massachusetts.

Weinhold, F. (1975a). *J. Chem. Phys.* **63**, 2479–2483.

Weinhold, F. (1975b). *J. Chem. Phys.* **63**, 2484–2487.

Weinhold, F. (1975c). *J. Chem. Phys.* **63**, 2488–2495.

Weinhold, F. (1975d). *J. Chem. Phys.* **63**, 2496–2501.

Weinhold, F. (1976a). *Phys. Today* **29**, 23–30.

Weinhold, F. (1976b). *Acc. Chem. Res.* **9**, 236–240.

Weinhold, F. (1976c). *J. Chem. Phys.* **65**, 559–564.

Wheeler, J. C. (1974). *J. Chem. Phys.* **61**, 4474–4489.

Wilcox, R. M. (1968). *Phys. Rev.* **174**, 624–629.

Wilson, E. B. (1936). *In* "A Commentary on the Scientific Writings of J. Willard Gibbs" (F. G. Donnan and A. Haas, eds.), Vol. I, pp. 19–59. Yale Univ. Press, New Haven, Connecticut.

A Model of the Liquid State. Three Phase Partition Functions

Mu Shik Jhon†
Institut de Biologie Physico Chimique,
University de Paris, Paris, France

Henry Eyring
Department of Chemistry, University of Utah
Salt Lake City, Utah

† Present address: Korea Advanced Institute of Science, P.O. Box 150, Cheong Ryang Ri, Seoul, Korea.

I. Earlier Models of the Liquid State

A. Introduction

Calculation of liquid properties from potential functions rapidly reaches the limits of feasibility as the molecules become complex. A treatment of complex systems can only succeed as we separate out the significant liquid structures from the inconsequential. It is with this in mind that this paper will cover the development of the model of the liquid state and its applications to many physical properties of the liquid. In an early book, "The Kinetic Theory of Liquids," Frenkel (1955) set forth his ideas of the liquid state. His considerations were qualitative. He never seriously undertook a quantitative treatment. Hirschfelder *et al.* (1954) provided an extensive coverage of early models. Barker (1963) discussed lattice theories of the liquid state, the cell model, the free volume model, and tunnel models. Following this, Eyring and Jhon (1969) considered recent versions of significant liquid structures.

Recently in the treatise "Physical Chemistry," Eyring *et al.* (1971) discussed several liquid theories. Here, we consider various liquid models, emphasizing significant liquid structures. A theory of the liquids is a most important element in a thorough understanding of many branches of physics and physical chemistry. Because of the complicated nature of the liquid state, i.e., strong interaction of particles and their degree of disorder, the theoretical developments have been far behind theories of the gaseous and solid state. The Helmholtz free energy is a function of volume, temperature, and composition and any other variables that can change the energy levels of the system.

If we take the Helmholtz free energy A as the ordinate of a hypersurface in a space in which the independent variables are the additional dimensions, we can then calculate all thermodynamic properties from the coordinates and derivatives of the hypersurface. Since the Helmholtz free energy A is expressed as

$$A = -kT \ln \sum_i \exp(-\varepsilon_i/kT) \tag{1.1}$$

all variables except temperature T influence the Helmholtz free energy through the energy levels ε_i. Here, in principle, all properties are calculable if the energy levels from quantum mechanics are provided. In practice there are difficulties.

The following classical mechanical approximation for A is often useful:

$$A = -kT \ln \left[(h^{3N} N!)^{-1} \int \cdot \, e^{-H/kT} \, dq_1 \cdots dp_{3N} \right]. \tag{1.2}$$

When the Hamiltonian H involves many nonseparable variables as it does as molecules approach each other in partial disarray in the liquid, the mathematical difficulties mount. The Hamiltonian contains a potential between neighbors and then a continuous network of interactions, extending through the whole of phase space. Since there are no exactly repeating unit cells present a model must be devised if calculations are to be possible.

Simulation techniques using a computer is an attractive way of getting experimental results for a liquid composed of a limited number of simple molecules. However, even the best computers may fail as the molecules become numerous or complicated (Eyring *et al.*, 1971). The Kirkwood superposition approximation, which assumes that the probability of a three-molecule configuration is the product of probabilities for the corresponding three configurations of pairs, leads to a solution by breaking an otherwise infinite chain of equations. Unfortunately, the calculated values are often disappointing (Kirkwood, 1935; Egelstaff, 1967; Eyring *et al.*, 1971). The Percus–Yevick (1958) equation and the perturbation method of Barker and Henderson (1968) are useful approximate solutions of this problem.

Another procedure is to develop a model of the liquid state that correctly reproduces the hypersurface A, and so gives all thermodynamic properties. If the model is correct, it should also provide the basis for the calculation of transport phenomena as well. As the maximum number of properties are calculated with increasing accuracy from a model, one gets the correct picture much as the calculations of the X-ray diffraction pattern from a model leads one to crystal structure.

We will consider briefly cell theories, hole theories, the tunnel theory, and some others with their recent improvements. Since the Helmholtz free energy A is related to the partition function for N molecules f_N, the model approach might be called "the partition function approach." Here,

$$A = -kT \ln f_N = -kT \ln \frac{1}{h^{3N}N!} \int \exp \frac{-H}{kT} dq_1 \cdots dp_N \tag{1.3}$$

where

$$H = \sum_{i=1}^{3N} \frac{p_i^2}{2m} + U. \tag{1.4}$$

In Eq. (1.4), U is the total potential energy, which depends on the coordinates, and p_i is the momentum of the ith particles. Integrating Eq. (1.3) over the moments p_i, one gets f_N for a real fluid.

$$f_N = \left(\frac{2\pi mkT}{h^2}\right)^{3/2N} \frac{1}{N!} \int \exp\left(-\frac{U}{kT}\right) dq_1 \cdots dq_N. \tag{1.5}$$

To obtain the Helmholtz free energy A from f_N, the integration of the term $\int \exp(-U/kT)\, dq_1 \cdots dq_N$ in Eq. (1.5) is the chief source of the difficulties. The direct evaluation of the partition function for the liquid state is impossible. Therefore, it is quite necessary to make approximations.

B. Cell Theory

According to the simple cell model (see Buehler *et al.*, 1951; Hirschfelder *et al.*, 1954), each molecule in a liquid or compressed gas spends much of its time restricted by its neighbors in a relatively confined space; i.e., the neighboring molecules form a "cell" in which the central molecule, the wanderer, moves. Therefore, the local environment in a liquid is not very different from that in a solid, except that the liquid molecule has only slightly more freedom than the solid. All cells are assumed alike and contain one molecule. The configurational partition function then is $f_c = 1/N! \int e^{-H/kT}\, dq_1 \cdots dp_{3N}$ and is given simply by

$$f_c = \left(\frac{1}{N!}\right) \sum{}^* v_f^N. \tag{1.6}$$

Here, the summation $\sum^*$ is taken over all arrangements of the N molecules in the N cells and the free volume v_f is the volume available to the center of a molecule in its cell. Since there are $N!$ different arrangements of N molecules in N cells, Eq. (1.6) becomes

$$f_c = v_f^N. \tag{1.7}$$

A simplification made is to assume that the free volume is approximated by a sphere of radius $(a - \alpha)$. Then,

$$v_f = \tfrac{4}{3}\pi(a - \alpha)^3. \tag{1.8}$$

Here a and α are the mean distance and the distance of closest approach between the center of neighboring molecules. Using the relation $p = -(\partial A/\partial V)_T$ and Eqs. (1.2)–(1.8), one can then evaluate the equation of state for a liquid. However, for liquids at ordinary pressures, it fails badly. This is due to the serious simplifying assumptions used for the highly artificial hard sphere potential function.

A much simpler approach was suggested by Eyring and Hirschfelder (1937) with qualitative success. Considering the effect of the attractive force, they concluded that U for many liquids can be represented by a formula, $U = -a/V$, where V is the molar volume of the liquid and a is a function of temperature. Clearly, a further improvement on the simple cell theory outlined above was made using the realistic potential function developed in successive papers by Lennard-Jones and Devonshire (1937). According to them, a molecule in a liquid spends much of its time in the cell surrounded by its neighbors at their equilibrium positions. To obtain a simplified calculation of the free volume, they assumed that z nearest neighbors can be treated as if they were uniformly smeared over a spherical surface of radius equal to the nearest neighbor distance a. Then, one obtains the partition functions for a system of N molecules

$$f_N = \frac{(2\pi mkT)^{3/2N}}{h^{3N}} \exp\left[-\frac{N\psi(0)}{2kT}\right] v_f^N [J(T)]^N \tag{1.9}$$

where $\psi(0)$ indicates the minimum in a pair interaction potential $\psi(r)$, $J(T)$ is the partition function for the internal degree of freedom of the molecule, and v_f is the free volume which is given as

$$v_f = \int_{\text{cell}} \exp\left[-\frac{\psi(r) - \psi(0)}{kT}\right] dp. \tag{1.10}$$

The Lennard-Jones (6-12) potential takes the form

$$\psi(r) = 4\varepsilon[(\sigma/r)^{12} - (\sigma/r)^6] \tag{1.11}$$

where ε and σ are the energy and distance parameters and are tabulated by Hirschfelder *et al.* (1954). In their calculations, the Lennard-Jones (6-12) potential was used. The cell geometry is shown in Fig. 1. Here the field $\psi(r)$ in which a molecule moves in its cell is the field of the surrounding molecules averaged over all directions. The area of the ring is $2\pi a^2 \sin\theta\, d\theta$ and therefore the number of smeared nearest neighbors in the ring is $(z/2) \sin\theta\, d\theta$.

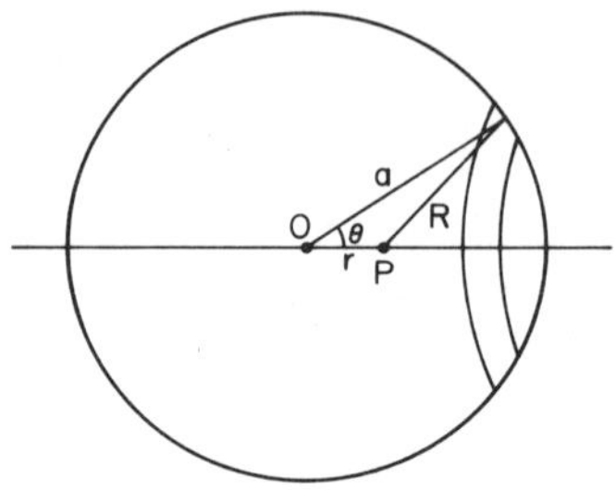

Fig. 1. Cell geometry in the Lennard-Jones and Devonshire model.

Hence,

$$\psi(r) = \int_0^\pi 4\varepsilon\left\{\left(\frac{\sigma}{R}\right)^{12} - \left(\frac{\sigma}{R}\right)^6\right\}\frac{z}{2}\sin\theta\, d\theta \tag{1.12}$$

where $R^2 = r^2 + a^2 - 2ar\cos\theta$. For r fixed, $2RdR = 2ar\sin\theta\, d\theta$ and

$$\psi(r) = \frac{2Z}{ar}\int_{a-r}^{a+r}\left\{\left(\frac{\sigma}{R}\right)^{12} - \left(\frac{\sigma}{R}\right)^6\right\} RdR. \tag{1.13}$$

Integrating the Eq. (1.13) and using the relation that

$$\psi(0) = 4z\varepsilon\{(\sigma/a)^{12} - (\sigma/a)^6\} \tag{1.14}$$

one obtains

$$\begin{aligned}\psi(r) - \psi(0) = {} & z\varepsilon\sigma^{12}\left\{\frac{1}{5ar}\left[\frac{1}{(a-r)^{10}} - \frac{1}{(a+r)^{10}}\right] - \frac{4}{a^{12}}\right\} \\ & + z\varepsilon\sigma^6\left\{\frac{1}{2ar}\left[\frac{1}{(a+r)^4} - \frac{1}{(a-r)^4}\right] - \frac{4}{a^6}\right\}.\end{aligned} \tag{1.15}$$

For a hexagonal close-packing structure, $z = 12$ and $V/N = a^3/\sqrt{2}$. Under these conditions, Eq. (1.15) can be rewritten as

$$\psi(r) - \psi(0) = z\varepsilon\left\{\frac{l(y)}{v^{*4}} - \frac{2m(y)}{v^{*2}}\right\} \tag{1.16}$$

where

$$l(y) = (1 + 12y + 25.2y^2 + 12y^3 + y^4)(1 - y)^{-10} - 1 \tag{1.17}$$

$$m(y) = (1 + y)(1 - y)^{-4} - 1 \tag{1.18}$$

and

$$y = (r/a)^2 = (r^*/a^*)^2, \qquad r^* = r/\sigma, \qquad a^* = a/\sigma, \qquad v^* = v/\sigma^3 = V/N\sigma^3.$$

Choosing the volume of the cell so that it equals V/N, the free volume v_f

Eq. (1.9) becomes

$$v_f = \int_0^{r_m} \exp\{-[\psi(r) - \psi(0)]/kT\} 4\pi r^2 \, dr = 2\pi a^3 g(v^*, T^*) \qquad (1.19)$$

where

$$g(v^*, T^*) = \int_0^{y_m} \sqrt{y} \exp\left\{-\frac{z}{T^*}\left[\frac{l(y)}{v^{*4}} - \frac{2m(y)}{v^{*2}}\right]\right\} dy \qquad (1.20)$$

and

$$y_m = \left(\frac{3}{4\pi\sqrt{2}}\right)^{2/3}. \qquad (1.21)$$

Then, using Eq. (1.9) and the relation $p = -(\partial A/\partial V)_T$, the pressure equation becomes

$$\frac{pV}{NkT} = -\frac{v}{2kT}\left(\frac{\partial\psi(0)}{\partial v}\right)_T + v\left(\frac{\partial \ln v_f}{\partial v}\right)_T. \qquad (1.22)$$

Here $v = a^3/\sqrt{2}$. The first term in the potential part of the pressure due to the potential energy of all the molecules placed at the centers of their cells and the second term is the thermal part of the pressure due to the motion of the molecules in their cells. Carrying out these differentiations, one gets

$$\frac{pV}{NkT} = 1 + \frac{2z}{T^*}\left[\frac{1}{v^{*4}}\left(1 + \frac{2g_l}{g}\right) - \frac{1}{v^{*2}}\left(1 + \frac{2g_m}{g}\right)\right] \qquad (1.23)$$

where g is given by Eq. (1.20) and

$$g_l(v^*, T^*) = \int_0^{y_m} l(y)\sqrt{y} \exp\left\{-\frac{z}{T^*}\left[\frac{l(y)}{v^{*4}} - \frac{2m(y)}{v^{*2}}\right]\right\} dy \qquad (1.24)$$

$$g_m(v^*, T^*) = \int_0^{y_m} m(y)\sqrt{y} \exp\left\{-\frac{z}{T^*}\left[\frac{l(y)}{v^{*4}} - \frac{2m(y)}{v^{*2}}\right]\right\} dy. \qquad (1.25)$$

Several authors have tabulated these integrals (Lennard-Jones and Devonshire, 1937; Prigogine and Raulier, 1942; Hill, 1947; Prigogine and Garikian, 1948). Following Lennard-Jones and Devonshire they considered only the effect of various shells beyond the first on the lattice energy but not on the free volume.

Wentorf *et al.* (1950) have carried out the most extensive calculations of the Lennard-Jones and Devonshire equation of state. They considered the effect of the three neighboring shells on both the lattice energy and on the free volume. The equation of state obtained is

$$\frac{pV}{NkT} = 1 - \frac{12}{T^*}\left[\frac{2.4090}{v^{*2}} - \frac{2.0219}{v^{*4}}\right] - \frac{48}{T^*}\left[\frac{1}{v^{*2}}\frac{G_M}{G} - \frac{1}{v^{*4}}\frac{G_L}{G}\right] \qquad (1.26)$$

and G, G_L, and G_M are integrals like g, g_l, and g_m except that $l(y)$ and $m(y)$ are replaced by the functions

$$L(y) = l(y) + \tfrac{1}{128}l(\tfrac{1}{2}y) + \tfrac{2}{729}l(\tfrac{1}{3}y) \tag{1.27}$$

$$M(y) = m(y) + \tfrac{1}{16}m(\tfrac{1}{2}y) + \tfrac{2}{27}m(\tfrac{1}{3}y). \tag{1.28}$$

The cell theory has been applied to quantum liquids by several authors (de Boer and Lunbeck, 1948; Pekeris, 1950; Levelt and Hurst, 1960; Hillier and Walkley, 1964; Henderson and Reed, 1964; Oden and Henderson, 1964; Reed and Henderson, 1964). The integrals in Eq. (1.2) must be replaced by a sum, and the energy levels must be calculated by solving Schrödinger's equation. Recently, Mullin (1971) derived the second-quantized cell model Hamiltonian for the model of a quantum liquid and a solid.

According to the experimental evidence the apparent number of nearest neighbors ordinarily decreases upon melting, and the liquid has a remarkable entropy increase upon melting. Hence, the cell theory has unavoidable serious defects. The thermodynamic properties calculated from the cell model are more nearly those of a superheated solid than a liquid. This is quite natural since the assumptions in the cell model are more characteristic of the solid state than of the liquid. The cell theory can be improved as we shall now see.

Without introducing the fixed cell geometry, Adams and Matheson(1972, 1973) examined an improved cell theory. The assumption made is that the cell centers are placed at the centers of irregularly packed hard spheres. Calculations using this theory and Lennard-Jones (6-12) potential gives significant improvement.

Another version is the cell cluster theory (de Boer, 1954; Barker, 1955). The unphysical assumptions of single occupancy of the cells and absence of correlations between molecules in neighboring cell are dropped. In successive corrections to the result of the cell model, factors due to collective motions in clusters of 2.3 ... cell are taken into account. However, this method has not been particularly successful. The consideration of clusters of two cells improve some properties, but make others worse (Weissmann and Mazo, 1962). This difficulty can be removed by considering high-order clusters. However, difficulties due to slow convergence are increased.

C. Hole or Free Volume Theory

The typical expansion of the substance upon melting (often about 12%) is accompanied by a decrease in the coordination number from 12 to about 10, with little change in distance of nearest neighbors. Also, the liquid has an entropy increase upon melting. The above facts are suggestive. One early explanation assumed the presence of a substantial amount of "communal entropy" in the liquid, not present in the solid but present in the gas (Hirschfelder *et al.*, 1937). The additional disorder and entropy are supplemented by "communal entropy" in the cell theory. Another explanation is as follows.

The cell theory can be improved if the number of the cells L is allowed to exceed the number of the molecules N so that some of the cells are unoccupied or there are "holes" in the liquid, as Eyring (1936; Cernuschi and Eyring, 1939) was led to postulate. The concept of the presence of holes and their more or less random distribution explains at least part of the entropy increase on melting. Now, if z_i is the coordination number of some particular molecule i then,

$$z_i = y_i z \tag{1.29}$$

where y_i is some number between zero and unity because of "holes" in the liquid. And if the random distribution of the molecules and holes is assumed, then $y_i = N/L = w/v$, where the volume per molecule is $v = V/N$ and volume per cell $w = V/L$. Then Eq. (1.9) becomes

$$f_N = \left[\frac{L!}{N!(L-N)!}\right]\frac{(2\pi mkT)^{(3/2)N}}{h^{3N}} \exp\left[\frac{-yNz\phi(0)}{2kT}\right](v_f)^N \tag{1.30}$$

where $\psi(0) = z\phi(0)$. In the limit of low densities, $L \geq N$ and $y_i \to 0$ so that $v_f \to w$

$$\begin{aligned} f_N &= \frac{L!}{N!(L-N)!}\frac{(2\pi mkT)^{(3/2)N}}{h^{3N}} \\ \cdot\, w^N &= \left(\frac{1}{[1-(N/L)]^L}\frac{(2\pi mkT)^{(3/2)N}}{h^{3N}}\right) \\ \cdot\, v^N &= \left(\frac{(2\pi mkT)^{3/2}}{h^3}\left(\frac{ev}{N}\right)^N\right). \end{aligned} \tag{1.31}$$

Hence, the hole or free volume models give the right low density limit. In order to obtain numerical results, the functional dependence of v_f on y must be known.

The early results from the Lennard-Jones and Devonshire theory are widely tabulated. If all the cells around the molecule are occupied; $y = 1$, $v_f \to v_f^0$ as tabulated by Wentorf *et al.* (1950). If all the surrounding cells are empty; $y = 0$, $v_f \to$ volume of the cell. Recently, Weissmann (1964) calculated v_f numerically in the presence of empty cells and found that determining L by minimizing the free energy, the number of holes is small. Hence, the hole or free volume model does not overcome the deficiencies of cell models except in the limiting case of the low density regions.

Actually, v_f does not depend on y in a simple way. However, many authors assumed the following linear dependency of v_f.

(a) Cernuschi and Eyring (1939)

$$v = v_f^0. \tag{1.32a}$$

(b) Ono (1947)

$$\ln v_f = y \ln v_f^0 + (1 - y) \ln q. \tag{1.32b}$$

(c) Peek and Hill (1950)

$$v_f = v_f(\bar{y}) \tag{1.32c}$$

where $\bar{y}$ is the average nonempty site around any molecule.

(d) Rowlinson and Curtiss (1951)

$$\ln v_f = \ln v(y) + (y - \bar{y})[(\partial/\partial_y) \ln v_f]_{y=\bar{y}}. \tag{1.32d}$$

(e) Henderson (1962)

$$v_f = yv_f^0 + (1 - y)q. \tag{1.32e}$$

The approximations of Cernuschi and Eyring and Ono are alike in that q is chosen as equal to the cell volume of the solid at the melting point. On the other hand, according to Peek and Hill and Rowlinson and Curtiss, the values of q are chosen so that the Helmholtz free energy is a minimum for each volume and temperature. The latter method is preferable theoretically, but too small a proportion of holes is predicted.

Grindlay (1961) has recently been able to evaluate the partition function explicitly in the quasichemical approximation without assuming a linear dependence of $\ln v_f$ on y. In his study, q was determined by minimizing the free energy. The results of all these calculations can be seen in Table I. Among them, Henderson's results are quite good and show a significant improvement over previous attempts.

Table I

CALCULATED VALUES OF THE CRITICAL POINT PROPERTIES[a]

Mean value for[b]	T_c^*	V_c^*	P_c^*	$P_c^* V_c^* / T_c^*$	y_c
Ne, Ar, N_2	1.28	3.15	0.119	0.293	~ 0.5
LJ & D	1.30	1.77	0.434	0.591	1.000
C & E	2.75	2.00	0.470	0.342	0.544
Ono	0.75	2.00	0.128	0.342	0.544
P & H	1.18	3.25	0.261	0.825	0.825
R & C	—	—	—	—	—
Grindlay	1.30	1.77	0.434	0.591	1.000
Henderson	1.41	3.39	0.139	0.333	0.321

[a] From Eyring and Jhon (1969, p. 21).

[b] Key to abbreviations: LJ & D, Lennard-Jones and Devonshire; C & E, Cernuschi and Eyring; P & H, Peek and Hill; R & C, Rowlinson and Curtiss.

Recently, a Kihara potential, instead of the Lennard-Jones (6-12) potential, was applied to the hole theory of Henderson by Nagata and Nishida (1969). The results show similar agreement with experiment to those in the original paper. Many modifications of the free volume theory of Lennard-Jones and Devonshire have been considered. However, the examples outlined above are representative. According to an early Kirkwood paper, the Lennard-Jones and Devonshire theory is a first approximation to a variational calculation of the free energy. However, the method is shown to be unsatisfactory by investigations of Mayer and Careri (1952), Dahler *et al.* (1956), and Barker (1955, 1956, 1957, 1958), among others.

Other approaches to correct the defects of the Lennard-Jones and Devonshire theory by direct calculation include giving up a spherical free volume, the correct calculation of the communal entropy, and the inclusion of the correlation with neighboring molecules. Buehler *et al.* (1951) performed the numerical integration over the dodecahedral cell for nonattracting rigid sphere molecules. They find a free volume correction factor of up to 1.4 in the critical region. The correction for the communal entropy has been made by Pople (1951) and by Janssens and Prigogine (1950) by considering the probability of finding two molecules in one cell. However, even in the dense gas region, it is found that double occupancy makes only a small contribution to the free energy. Later, Weissmann and Mazo (1962) studied the double occupancy effect in much more detail than formerly, and their results are in the right direction for the liquid density ranges. Unfortunately, the corrections for the entropy are too small.

These facts point out the need for such further refinements as molecular correlation terms. In a series of papers, similar considerations were undertaken by deBoer and Cohen and colleagues (de Boer, 1954; Cohen *et al.*, 1955, 1957; Salsburg *et al.*, 1957; Cohen and Rethmeier, 1958). In their method, the multiple occupancy and the correlation corrections are both included in one term, the cell cluster integral. This seems to not improve significantly on Barker's procedure where the two effects are considered separately. However, the slow convergence of the computation for the higher occupied cells is again a problem.

D. Tunnel Theory

Recently, Barker (1960) developed the tunnel model for fluids. In his model, the liquid is disordered in a way more fundamental than the cell or hole model considers. In cell theory, the whole fluid is divided into subsystems sufficiently simple to permit calculation of their properties. Barker maintains this idea and chooses the subsystems to be lines of molecules moving almost one dimensionally in tunnels, the walls of the tunnel being formed by neighboring lines of molecules. Since each subsystem contains a large number of molecules, the problem of communal entropy does not arise. The subsystems are approximately one dimensional, thus the partition function can be evaluated easily. It is assumed that the volume V is divided into K tunnels each containing N molecules and that the longitudinal and transverse motions of the molecules in a given tunnel are separable. Then, the partition function of the fluid becomes

$$f_N = [f_{\mathrm{M}}(1)]^K [f(2)]^N \tag{1.33}$$

where $f_{\mathrm{M}}(1)$ is the one-dimensional partition function for longitudinal motion and $f(2)$ is the two-dimensional partition function for the motion across the tunnel (i.e., transverse motion). Here, $f_{\mathrm{M}}(1)$ can be expressed as

$$f_{\mathrm{M}}(1) = \left(\frac{(2\pi mkT)^{1/2}}{h}\right)^M \left[\frac{1}{M!}\int_0^{lM} \cdots \int_0^{lM} \exp\left[-\frac{u(x)}{kT}\right] dx_1 \cdots dx_M\right] \tag{1.34}$$

where $u(x)$ is the potential energy of the one-dimensional system of M molecules in the length lM. When there are K such tunnels in a system, the total partition function will be f_{M}^K.

The partition function for the motion across the tunnel can be evaluated in terms of a "free area," a_{f}, analogous to the "free volume" in the cell model. Then we have

$$f(2)^N = \frac{(2\pi mkT)^{(3/2)N}}{h^{3N}} \left[a_{\mathrm{f}} \exp - \left(\frac{\psi(0)}{2kT}\right)\right]^N \tag{1.35}$$

where

$$a_{\mathrm{f}} = \int \exp\{-(\psi(r) - \psi(0)/kT\} \, dr. \tag{1.36}$$

In Eq. (1.36), $\psi(r)$ is the potential energy of interaction of a given molecule at a distance r from the axis of its tunnel, and this is summed.

If the tunnels are assumed to be close-packed, the values of $\psi(0)$ and a_{f} in Eq. (1.35) can be obtained.

$$V/N = \tfrac{1}{2}\sqrt{3}\, a^2 l \tag{1.37}$$

$$\psi(0)/kT = 4.652C - 7.925B \tag{1.38}$$

and

$$a_{\mathrm{f}} = a^2 S \tag{1.39}$$

where

$$S = 2\pi \int_0^{0.52} \exp - \left(\frac{\psi(y) - \psi(0)}{kT}\right) y \, dy \tag{1.40}$$

where $y = r/a$ and a is the distance between the center of neighboring tunnels. The one-dimensional partition function $f_M(1)$ can be written as

$$-\frac{1}{N}\ln\left[f_M\left(\frac{(2\pi mkT)^{1/2}}{h}\right)^{-M}\right] = \frac{A_1^*}{RT} + \frac{4}{T^*}\left[0.0003\left(\frac{\sigma}{l}\right)^{12} - 0.0173\left(\frac{\sigma}{l}\right)^{6}\right]. \tag{1.41}$$

The last term is the Helmholtz free energy due to the nonnearest neighbor interaction; the second term due to the nearest neighbor interaction is

$$A_1^* = -\frac{p_1 l}{kT} - \ln\left[\int_0^\infty \exp\{-[u(x) + p_1(x)]/kT\}\, dx\right]. \tag{1.42}$$

Here p_1 is the one-dimensional pressure related to the intermolecular distance l by Eq. (1.43)

$$l = \int_0^\infty x \exp\{-[u(x) + p_1(x)]/kT\}\, dx \bigg/ \int_0^\infty \exp\{-[u(x) + p_1(x)]/kT\}\, dx. \tag{1.43}$$

With the use of the relation, $A = -kT \ln f_N$ and Eqs. (1.37) to (1.43), one obtains the equation of state

$$pV/NkT = \frac{1}{T^*}\left(\frac{V_0}{V}\right)^4\left(20.938 + 9\frac{S_c}{S}\right) - \frac{1}{T^*}\left(\frac{V_0}{V}\right)^2\left(23.880 + 6\frac{S_B}{S}\right) + \frac{0.3497}{w}\left(\frac{V_0}{V}\right)^{-1/3} + \frac{2}{3}. \tag{1.44}$$

In Eq. (1.44), the relation $l = a = \sigma(2V/\sqrt{3}\,V_0)^{1/3}$ is introduced, where

$$S_B = \frac{\partial S}{\partial B}, \qquad S_c = -\frac{\partial S}{\partial C}, \qquad w = \frac{kT}{P_1\sigma}, \qquad \text{and} \qquad V_0 = N\sigma^3.$$

S, S_B, S_c, and w have been tabulated by Barker (1961).

In Table II, a comparison between theory and experiment of some properties at $kT/\varepsilon = 0.70$ are given and show good agreement for certain properties for argon. The tunnel theory is accordingly a theory of a liquid rather than a solid. However, the value of the entropy gives poor agreement. The difficulty is with the assumption that the molecules in different tunnels move indepen-

Table II

THEORETICAL AND EXPERIMENTAL PROPERTIES AT $kT/\varepsilon = 0.70$[a]

	Reduced volume (V/V_0)	Reduced excess energy ($E^E/N\varepsilon$)	Reduced excess entropy (S^E/Nk)
Tunnel theory	1.184	−5.9	−4.8
Liquid argon	1.181	−5.96	−3.64
LJ & D theory	1.037	−7.32	−5.51
Solid argon	1.035	−7.14	−5.33

[a] From Eyring and Jhon (1969, p. 25).

dently, thus the model underestimates the volume of configuration space associated with these structures. Barker (1962) and Chung and Dahler (1964) have shown how these assumptions should be corrected.

A treatment of correlation effects analogous to those considered in the cell model should improve liquid theory. Oden and Henderson (1964) have formulated a quantum tunnel model and obtained some results for quantum hard spheres. Very recently, the extension of Barker's tunnel model to binary mixtures of simple liquid (Malijevský, 1971) has been carried out. The agreement between theory and experiment is better than that of the Prigogine average potential model and also better than the van der Waal approximation in the method of pseudopotentials.

E. OTHER LIQUID MODELS

A worm model of the liquid state, based on Barker's tunnel model, which assumes that the molecular interactions are pair additive and are given by the Lennard-Jones (6-12) potential has been developed by Chung and Dahler (1964, 1965a,b). The initial calculations for the limited high density regions have been extended to lower densities. The model shows a significant improvement over the cell theory for a limited range at high pressures.

An extension or modified version of the model of the liquid based on the lattice model of Guggenheim (1952) is studied by Bell and Lavis (1970). They consider a bonded fluid model on a plane triangular lattice. Each molecule has three bonding directions at angles that are 120° to each other with two possible orientations in each where the bonding directions point to three of the six nearest neighbor sites. Critical points and densities of liquid at various temperature and pressures are evaluated. It is interesting to note that at high pressures (i.e., no vacant sites) the model is equivalent to an Ising antiferromagnet on a triangular lattice.

II. The Significant Liquid Structures Model

A. Experimental and Philosophical Basis of the Model

1. Introduction

In Section I we discussed other models of the liquid, all of which need improvement to explain the nature of the liquid state. Here, the philosophical basis of the significant structure theory (SST) (Eyring *et al.*, 1958a; Eyring and Jhon, 1969; Jhon and Eyring, 1971) is discussed. There is a great accumulation of data on all kinds of liquids that a reliable theory must explain. The problem is to find an appropriate model of the liquid state in terms of the "normal coordinates." This should lead to a partition function $f = \sum_i \exp(-\varepsilon_i/kT)$ where the energy ε_i of the ith level for the system depends on such variables as volume, composition, and the potential fields that may be operating on the system. All the thermodynamic properties may then be calculated from the Helmholtz free energy A, where A takes the form $A = -kT \ln f$.

This procedure is very similar to the well-known methods for comparison of the experimental diffraction pattern to a diffraction pattern calculated from a model. The required ingenuity lies in selecting the right model for making the comparison. One then revises the model if necessary until the agreement is satisfactory.

This model should also be physically believable and be useful in calculating other properties and in guiding further experimentation. This is the significant structure theory program. It may be instructive to others to follow the thread of intuition in this development with the idea of improving on it where necessary.

2. Experiment and Philosophical Guidelines

The typical "normal liquid," such as argon, expands about 12% in melting. Since X rays (Eisenstein and Gingrich, 1942) indicate the nearest neighbor distance is essentially the same as for the solid, a liquid near the melting point must approximate a defective close-packed lattice with every eighth lattice site empty. For the lack of a better term for the empty sites they have been called "fluidized vacancies." A fluidized vacancy can exist in such an equilibrium situation only as long as the excess local kinetic energy density exerts an outward pressure that balances the pressure due to volume-dependent potential energy density plus any external pressure tending to make the vacancy collapse. This picture explains why a solid melts.

As the temperature is raised, the kinetic energy density (the pressure promoting expansion) rises in various localities while the counteracting potential energy density opposing expansion decreases with the temperature rise. When the critical balance between kinetic and potential energy densities is reached, holes enter cooperatively, and the solid moves to a new equilibrium situation, which is a solid studded with fluidized vacancies corresponding to a new balance between the opposing forces. Any one who has watched a space ship launched has observed a parallel phenomenon. Upon ignition, the rocket remains temporarily in its cradle. When enough fuel has burned so that the upward thrust due to the burning flame overcomes the gravitational attraction the rocket moves toward a new equilibrium situation much as a solid melts.

X-Ray diffraction reveals the existence of short-range order in the liquid state, but long-range order has disappeared. The existence of fluidized vacancies of varying shapes accounts for the disappearance of long-range order. What characterizes the distribution in size of fluidized vacancies? They tend to be of molecular size, since to be smaller would be like buying a shoe that is too small. Too much energy has been paid for the return in entropy. The same will be true if they are larger than molecular size, since too much will have been paid in energy for space that cannot be effectively used by a molecule. Thus, it is quite natural to expect the distribution in the volumes of fluidized vacancies to cluster about a mean volume that will accommodate a molecule. In this connection theoretical arguments will be discussed later. It is quite interesting to relate the concentration of fluidized vacancies in the liquid phase to the concentration of molecules in the vapor phase. If a molecule is taken from a condensed phase into the dilute vapor leaving a vacancy behind, all bonds joining the vaporized molecule to others are broken. If molecules are simultaneously vaporized without leaving any vacancies, again all bonds are broken, but since each bond belongs to two molecules the energy of vaporization is equal to only half of the sum of the energy of bonds joining one molecule to all near and distant neighbors. Therefore, it follows that the energy to make a vacancy is just the energy of vaporizing a molecule.

We may also relate the entropy associated with a vapor molecule to the entropy associated with a fluidized vacancy. A fluidized vacancy transforms three vibrational degrees of freedom of the liquid into three translations due to neighboring molecules cooperatively moving the vacancy about much the way a molecule moves about in the vapor. In fact, two fluidized vacancies collide and separate or combine somewhat the way two molecules do (see Fig. 2).

Since both the entropy and energy are almost the same for molecules and for fluidized vacancies, we expect that a cubic centimeter of vapor containing the same number of molecules as the number of fluidized vacancies in a cubic centimeter of liquid. Hence, the mean density of liquid and vapor is nearly equal to half the density of the liquid at the melting point except for a

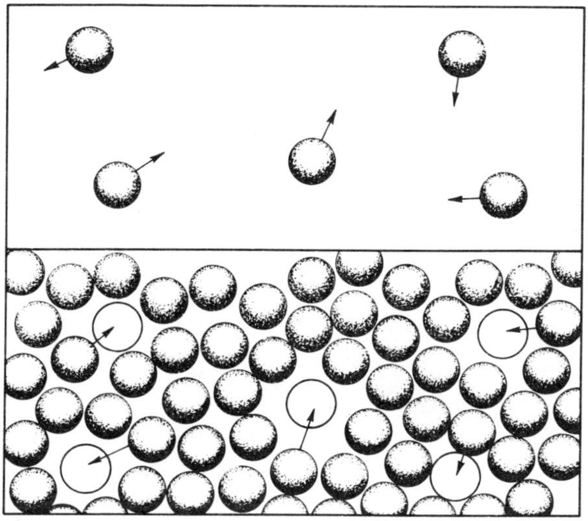

Fig. 2. Vacancies in a liquid behave like molecules in a gas. In a liquid, vacancies move among molecules. In a gas, molecules move among vacancies. Figure adapted from Moore (1962).

linear drift downward with temperature due to the expansion of the liquid quasi-lattice. Thus, the significant liquid structure model readily explains the law of rectilinear diameter of Caillelet and Mathias (1886, 1887).

B. Formulation of the Model; the Partition Function for a Simple Liquid

In view of the previous discussion, we can construct the partition function. Since the partition functions for solidlike and gaslike molecules are well known, we must determine in what proportion they occur in a liquid. Also, it is explained above that fluidized vacancies are expected to have a mean volume equal to that of the molecules in which they are found. Consequently, the number of moles of fluidized vacancies per mole of liquid is $(V - V_s)/V_s$, where V and V_s are the molal volumes of liquid and solid, respectively. Since, on the average, a molecule will encounter a near neighbor on V_s/V of its vibrations, that fraction of its trips it will act like a solid and the remaining number of trips it will fall freely and be gaslike. According we treat the fraction V_s/V of the degrees of freedom as solidlike and the remaining $(V - V_s)/V$ as gaslike.

Now, besides replacing three vibrational degrees of freedom with three translations, a fluidized vacancy introduces positional degeneracy of the solidlike molecule. Accordingly, we can construct the molal partition function, f_N of the system of N (Avogadro's number) molecules as follows:

$$f_N = (f_s)^{N(V_s/V)}(f_g)^{N[(V-V_s)/V]}. \tag{2.1}$$

If we assume that an Einstein oscillator is an adequate representation of the lattice vibrational degree of freedom of the solidlike molecules for all but quantum liquid at the temperature of interest, we can write the solidlike partition function f_s for a monatomic liquid such as argon as

$$f_s = \exp(E_s/RT) \left[(1 - e^{-\theta/T})^{-3} + n_h \exp(-\varepsilon_h/kT)\frac{(2\pi mkT)^{1/2}}{h} l(1 - e^{-\theta/T})^{-2}\right]. \tag{2.2}$$

Here E_s is the energy of sublimation at the melting point and θ is the Einstein characteristic temperature. n_h is the number of neighboring vacancies a molecule has a chance to preempt and ε_h is the energy needed to deny the position to competing neighbor molecules. Thus $n_h \exp(-\varepsilon_h/kT)$ is the number of degenerate positions possessed by a solidlike molecule and

$$(\exp E_s/RT)\frac{(2\pi mkT)^{1/2}}{h} l(1 - e^{-\theta/T})^{-2} \tag{2.3}$$

is the addition to the partition function provided by such a degenerate position. The passage way between the two lattice points is a translation that accounts for the translational degree of freedom in Eq. (2.3), while the other two degrees of freedom are vibrations. An estimate of l is required. Since a molecule shears past one of the corners of the hexagon of positions made by the six neighbor molecules in the plane normal to the translational molecules with the other five corners of the hexagon articulating to allow the motion, we should have $l = \lambda/6$ for the displacement, with λ approximating a molecular diameter. This agrees with the value of $l = 0.634$ Å used for argon. The partition function for the gaslike molecules f_g is

$$f_g^{N[(V-V_s)/V]} = \left(\frac{(2\pi mkT)^{3/2}}{h^3}(V - V_s)\right)^{N[(V-V_s)/V]} \bigg/ \left(N\frac{(V - V_s)}{V}\right)!$$

$$= \left(\frac{(2\pi mkT)^{3/2}}{h^3}\frac{eV}{N}\right)^{[(V-V_s)/V]}. \tag{2.4}$$

Here, the gaslike molecule moves freely in the empty space $(V - V_s)$, and, as usual for gaslike behavior, we must divide by the symmetry number, i.e., the factorial of the number of gaslike molecules. Here, m is the mass of the molecules. We require values for n_h and ε_h.

Since n_h should be proportional to the number of vacancies, we write

$$n_h = n[(V - V_s)/V_s]. \tag{2.5}$$

The energy, ε_h should be inversely proportional to the number of moles of vacancies $(V - V_s)/V_s$ and directly proportional to the energy of sublimation E_s. Accordingly we write

$$\varepsilon_h = a[E_s V_s/(V - V_s)]. \tag{2.6}$$

Using Eqs. (2.2)–(2.6) and substituting into Eq. (2.1), we obtain

$$f_N = \exp\frac{E_s}{RT}\left((1 - e^{-\theta/T})^{-3} + n_h \exp - \frac{aE_s V_s}{(V - V_s)RT}\frac{(2\pi mkT)^{1/2}}{h} l(1 - e^{-\theta/T})^{-2}\right)^{NV_s/V} \times \left[\frac{(2\pi mkT)^{3/2}}{h^3}\frac{eV}{N}\right]^{[N(V - V_s)/V]}. \tag{2.7}$$

For a simple liquid, the value of E_s, θ, and V_s are obtained from the solid data although these properties of the solid could be calculated using an appropriate potential function. Their volume dependence can often be neglected. Methods of evaluating n and a have been given (Eyring and Jhon, 1969). The theory works without adjustable parameters for simple liquids. At the melting point, the liquid approximates a lattice so we can write for the number of vacancies adjoining a molecule the expression

$$n\frac{V - V_s}{V_s} = z\frac{V_m - V_s}{V_m} = z\frac{V_s}{V_m}\left(\frac{V_m - V_s}{V_s}\right). \tag{2.8}$$

Here V_m is the liquid molar volume at the melting point. Hence, $n = z(V_s/V_m) = 12 \times 1/1.12 = 10.7$ since the z in a close packed lattice is 12. Now, we can evaluate the proportionality factor a. At the melting point, we have (Eyring and Jhon, 1969)

$$\frac{aE_s V_s}{V_m - V_s} = \frac{(n - 1)}{z}\frac{1}{2}(3/2kT)N = \frac{(n - 1)}{z}\frac{1}{2}E_s\left(\frac{V_m - V_s}{V_m}\right). \tag{2.9}$$

Here the average energy the $(n - 1)$ competing neighbors for a vacant site can put into the vacancy at the melting point is $(n - 1)/z\frac{1}{2}((3/2)kT)$. The factor $1/z$ expresses the fact that only one-zth of the time will a molecule be headed toward the vacancy. The factor $\frac{1}{2}$ arises because a trip between the two sites causes equal sharing of the kinetic energy $(3/2)kT$. Since there is no change in temperature in melting, the kinetic energy is unchanged and the potential energy increase is $E_s(V - V_s)/V$ due to the empty lattice sites. These sites are prevented from collapsing by the kinetic energy $(3/2)RT$ so that $E_s(V - V_s)/V = (3/2)RT$ and Eqs. (2.8) and (2.9) give $a = 0.0052$ and $n = 10.7$, as compared with the best-fit values of $a = 0.00534$ and $n = 10.8$ arrived at during the early stages of the significant structure approach (Fuller *et al.*, 1959).

Some of the properties of the monatomic liquid are shown in Table III. Avoiding the introduction of adjustable parameters, the results show excellent agreements between calculated and experimental values. For more complicated liquid, we can still use the partition function

$$f_N = [f_s]^{N(V_s/V)}[f_g]^{N[(V-V_s)/V]}. \tag{2.10}$$

The evaluation of the parameters requires a more extended analysis to obtain them *a priori* for molecules that rotate. Good estimates of E_s, V_s, and θ can be obtained from the solid state. It is very interesting that the ratio n/V_s and aE_s/RT_m is constant for certain systems (Faerber *et al.*, 1972). These relations can be useful in obtaining n and a for complicated systems from the known values for argon. Alternatively, after a suitable model that fixes the expressions of f_s and f_g is chosen, the parameters may be chosen to fit the observed thermodynamic properties.

Table III

VAPOR PRESSURES, MOLAR VOLUMES, AND ENTROPY OF VAPORIZATION[a]

Temperature (°K)	P_{calc} (atm.)	P_{obs} (atm.)	V_{calc} (cm^3)	V_{obs} (cm^3)	ΔS_{calc} (e.u.)	ΔS_{obs} (e.u.)
			Argon			
83.96 (m.p)	0.6874	0.6739	28.84	28.03	20.07	19.43
87.49 (b.p)	1.040	1.000	29.36	28.69	18.90	18.65
97.76	2.883	2.682	31.04	30.15	15.92	—
			Krypton			
116.0 (m.p)	0.7605	0.7220	34.31	34.13	20.14	—
119.93 (b.p)	1.0660	1.000	34.90	—	19.15	17.99
			Xenon			
161.3 (m.p)	0.8372	0.804	42.24	42.68	20.25	—
165.1 (b.p)	1.0623	1.000	42.84	—	19.50	18.29

E_s, θ, and V_s values used

	Ar	Kr	Xe
E_s (cal/mole)	1888.6	2740	3897.7
V_s (cm^3/mole)	24.98	29.6	36.5
θ (°K)	60.0	45.0	39.2

[a] From Eyring and Jhon (1969, p. 34).

One of the basic ideas of the model is that all species present in the vapor are copied in the liquid as vacancies. These concepts lead to the following expression for the mean values of a liquid property X:

$$X = X_s \frac{V_s}{V} + X_g \frac{V - V_s}{V} \tag{2.11}$$

where X_s and X_g are the values of the property for the solidlike part of the liquid and in the vapor state, respectively. The property X can be the specific heat, the viscosity, the thermal conductivity, the surface tension, or the mean dipole moment of the liquid, etc. The early estimate of Walter and Eyring (1941) of the specific heat, based on the concept, is quoted frequently. Including the translational degree of freedom in f_s [see Eqs. (2.2) and (2.7)] brings the calculated C_v close to recent experimental values (Rowlinson, 1959).

C. Improvement and Refinement of the Model

Obviously, a successful model should not be in conflict with the Monte Carlo (Wood, 1968) and molecular dynamic calculations (Alder and Wainwright, 1957; Wainwright and Alder, 1958) of the properties of systems of hard spheres which form the simplest of liquids. In the usual investigation of the model, an Einstein oscillator is used for the solidlike degrees of freedom. Henderson (1963) replaced the Einstein oscillator by the Lennard-Jones and Devonshire cell models and also formulated the partition function for a rigid sphere that was applied with success. If a Lennard-Jones and Devonshire partition function is used as the solidlike partition function, one obtains

$$f_s^N = \left[\left(\frac{2\pi mkT}{h^2}\right)^{3/2} v_f\right]^N \exp\left\{\frac{-N\psi(0)}{2kT}\right\}. \tag{2.12}$$

Here, the notation except for the free volume v_f has been defined previously. If the Lennard-Jones (6-12) potential is assumed, then

$$\psi(0) = z\varepsilon\left[\left(\frac{1}{w^{*4}}\right) - \left(\frac{2.4}{w^{*2}}\right)\right] \tag{2.13}$$

where $w^* = V_s/N\alpha^3$ and the effect of nonnearest neighbors as well as of the z nearest neighbors has been taken into account. Substituting Eqs. (2.3)–(2.5) and (2.12) into Eq. (2.1) gives

$$f_N = \left(\frac{2\pi mkT}{h^2}\right)^{(3/2)N} \left[\exp\left(-\frac{\psi(0)}{2RT}\right)\left(1 + n\frac{V - V_s}{V}\exp(-\varepsilon_h/RT)\right)v_f\right]^{N(V_s/V)} \times \left(\frac{eV}{N}\right)^{N[(V-V_s)/V]}. \tag{2.14}$$

For the system of rigid spheres, $\psi(0) = 0$ and $\varepsilon_h = 0$. By means of machine calculations (Wood and Jacobsen, 1957; Alder and Wainwright, 1957), a transition from a solid phase to a fluid phase at about $V = 1.5\ V_0$ is found, where $V_0 = N\alpha^3/\sqrt{2}$ in the close packed volume of the system. Accordingly, the value $V_s = 1.5\ V_0$ was used in this calculation. The values of v_f for $V_s = 1.5\ V_0$ were obtained by interpolating the results of Buehler *et al.* (1951) and also $n = 12$ was taken. Then, one obtains the partition function for a system of hard spheres

$$f_N = \left\{\left(\frac{2\pi mkT}{h^2}\right)^{3/2} \frac{eV}{N}\right\}^N \{[0.00967(8 - 11(V_0/V)]^{(3/2)N(V_0/V)}\}. \quad (2.15)$$

A comparison of calculations with experiment is given in Fig. 3. In this

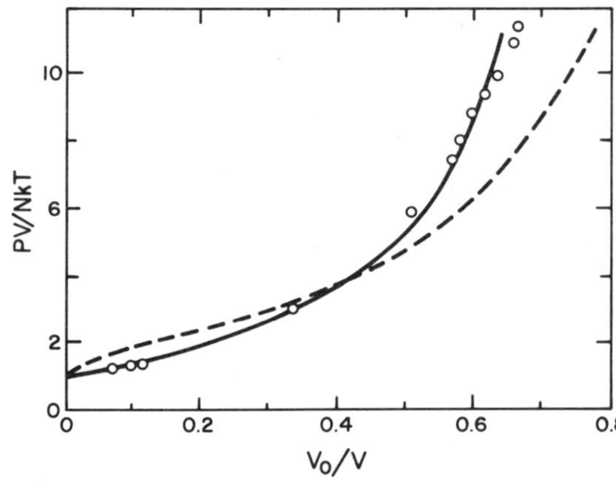

Fig. 3. Compressibility factor for a system of hard spheres. (——) The results of the present calculations; (- - -) the results of the cell model; (○) the results of the machine calculations of Alder and Wainwright.

calculation, the following expressions for the relation $p = -(\partial A/\partial V)_T$ were used:

$$\frac{pV}{NkT} = 1 - 3/2(V_0/V)\ \ln\{0.00967[8 - 11(V_0/V]\} + 3/2$$
$$+ 11(V_0/V)^2/[8 - 11(V_0/V)]. \quad (2.16)$$

The theory provides excellent agreement with the corresponding results of the machine calculations.

Recently, work was begun in an effort to understand the nature of melting of a hard sphere system, but results are not yet available. Another effort is underway to improve the partition function for a dense gas or the liquid under high pressure. The Einstein oscillator is modified to provide for the gradual transition from a parabolic to a square well potential with success (Ree *et al.*, 1962). To formulate the solidlike partition function, we assumed that the Einstein oscillator satisfies the condition $l\theta \geq T$, where l is the vibrational quantum number of the energy level at which the solid molecule sublimes, that is, $l = E_s/3R\theta$. However, for the high temperature region of the dense gas, the condition $l\theta \gg T$ no longer holds. This is particularly true for the substance having a low value of E_s such as argon and neon. Here, a more general partition function has been introduced:

$$f_s = \exp(E_s/RT)\left\{\frac{1 - e^{-l\theta/T}}{1 - e^{-\theta/T}} + e^{-l\theta/T}\left[\left(\frac{2\pi mkT}{h^2}\right)^{1/2}\left(V_s^{1/3} - \left(\frac{b}{4}\right)^{1/3}\right)\right]\right\}^3. \tag{2.17}$$

Here b is the van der Waal's constant, $b/4$ being the net volume of the molecule. In addition to the above correction, the modification of the value of V_s is also considered:

$$V_s = V_s(1 - K\,\Delta p) \tag{2.18}$$

where K is the compressibility of the solid phase. The results based on Eqs. (2.17) and (2.18) are compared with that of Wentorf *et al.* (1950), Henderson (1962), and Kirkwood *et al.* (1952) in Fig. 4. One sees that the significant structure approach agrees better than any of the others plotted in Fig. 4.

In general, it is known that the calculated critical pressure values in the significant structure model are higher than those found experimentally. This

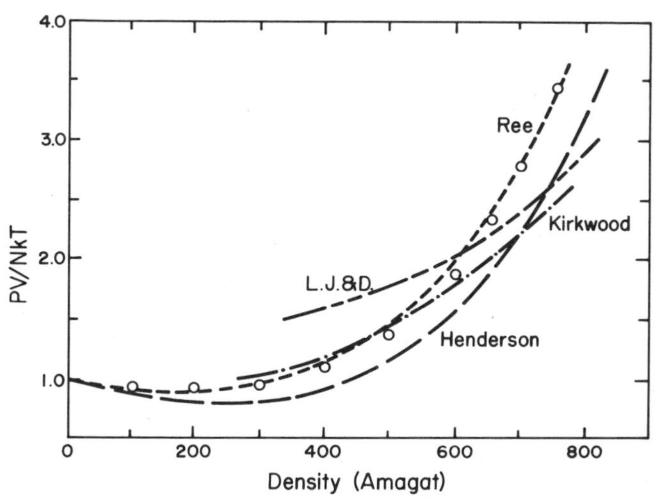

Fig. 4. Compressibility factor of argon at 0°C. The points represent the experimental values.

is due, at least in part, to the neglect of clustering of vacancies in the partition function. As an initial trial, the modest correction that includes a dimer term (Grosh *et al.*, 1967) improves the result. However, a more systematic rigorous attempt to take account of clustering should solve this defect. One of the important things to justify or understand in the model is the distribution of fluidized vacancies. In previous discussions, we have assumed that the most probable volume of a fluidized vacancy is the volume of the solid molecule. This has been discussed recently (Choi *et al.*, 1977).

The following inverse Gaussian distribution function is used for the vacancy size distribution:

$$\phi(j) = [\lambda/(2\pi j)^3]^{1/2} \exp[-\lambda(j-\mu)^2/2\mu^2 j] \tag{2.19}$$

where $j = V_h/V_s$, V_h is the volume of a vacancy, $\phi(j)$ is the probability of having a j-sized vacancy, and λ and μ are the parameters to be determined. Equation (2.19) is included explicitly in the partition function for a simple liquid such as argon.

$$f_N = \left\{\frac{e^{E_s/RT}}{(1-e^{-\theta/T})^3} \times \left[1 + n\left(\frac{V-V_s}{V_s}\right)\left(\sum_j j\phi(j)\exp\left\{-\frac{aE_s V_s}{(V-V_s)jRT}\right\}\Big/\sum_j \phi(j)\right)\right]\right\}^{N(V_s/V)} \times \left\{\frac{(2\pi mkT)^{3/2}}{h^3}\frac{eV}{N}\right\}^{N[(V-V_s)/V]} \tag{2.20}$$

Assuming that $\phi(j)$ has a maximum values at $j = 1$, and using the relation $[\partial\phi(j)/\partial j]_{j=1} = 0$, one obtains $\lambda = 3\mu^2/(\mu^2 - 1)$. And with the use of the relation $\partial A/\partial\mu = 0$, where $A = -kT \ln f_N$, μ is found to be 1.0005. This indicates that the vacancy size distribution is nearly a δ function, which support one of the assumptions of the model for simple liquids (see Table IV). Extensive calculations for various properties in many systems are in progress.

D. Comparison with Other Models

We now consider the reduced equation of state for various models and compare the thermodynamic properties for simple liquids. The partition function and the equation of state for other models were discussed in Section I. According to the significant structure theory, the partition function (Reed

Table IV

VALUES OF $\phi(j)$ FOR VARIOUS j[a]

j	$\phi(j)$
0.0–0.7	0
0.8	5.4×10^{-32}
0.9	1.2×10^{-5}
1.0	21.9
1.1	2.6×10^{-5}
1.2	4×10^{-21}
1.3	1.6×10^{-44}
1.47	0

[a] From Choi *et al.* (1977).

and Henderson, 1964) for a simple liquid is

$$f_N = \left[e^{E_s/RT}\left(\frac{T}{\theta}\right)^3 (1 + n_h e^{-\varepsilon_h/RT}) J(T)\right]^{N/X}$$
$$\times \left[\left(\frac{2\pi mkT}{h^2}\right)^{3/2} (V - V_s) J(T)\right]^{N(1-1/X)} \left[\frac{N(V - V_s)}{V}!\right]^{-1}. \quad (2.21)$$

Here $X = V/V_s$. Equation (2.21) is the familiar expression for the significant liquid structure partition function. Here, the following classical approximation to the Einstein partition function was used:

$$\frac{1}{(1 - e^{-\theta/T})^3} \simeq \left(\frac{T}{\theta}\right)^3 \quad (2.22)$$

where $\theta < T$ and $J(T)$ is the partition function for the internal degree of freedom. In Eq. (2.21), the configurational energy E_s can be written (Pekeris, 1950; Mullin, 1971) in terms of the Lennard-Jones (6-12) potential;

$$E_s = \frac{z}{2} N\varepsilon \left[2.4090\left(\frac{V_s}{N\sigma^3}\right)^{-2} - 1.0109\left(\frac{V_s}{N\sigma^3}\right)^{-4}\right]. \quad (2.23)$$

At the nearest neighbor distance a for the solid, the interaction energy $\psi(r)$ given by Eq. (1.11) has its minimum value. Then, we have the relation

$$a = 2^{1/6}\sigma. \quad (2.24)$$

Using Eq. (2.24) for hexagonal close packing, one obtains

$$V_s = Na^3/\sqrt{2} = N\sigma^3. \quad (2.25)$$

The Einstein characteristic temperature θ in the partition function can be related as follows:

$$\theta = \frac{h\nu}{k} = \frac{h}{k}\frac{1}{2\pi}\left(\frac{b}{m}\right)^{1/2}. \tag{2.26}$$

In this derivation, a simple harmonic oscillator has been assumed and the force constant b is approximated using the Lennard-Jones (6-12) potential as

$$b(r^2/2) \simeq z(\psi(r) - \psi(a)). \tag{2.27}$$

Then (Adams and Matheson, 1972)

$$b = 2z\varepsilon\left[22.106\left(\frac{N\sigma^3}{V_s}\right)^4 - 10.559\left(\frac{N\sigma^3}{V_s}\right)^2\right]\left[\frac{1}{2^{1/3}\sigma^2}\left(\frac{N\sigma^3}{V_s}\right)^{2/3}\right]. \tag{2.28}$$

The values of n and a for simple liquid are taken from Eqs. (2.8) and (2.9).

Introducing Eqs. (2.5)–(2.6), (2.8)–(2.9), and (2.22)–(2.25) into Eq. (2.21) and applying Stirling's approximation to $[N(V - V_s)/V]!$, one obtains

$$f_N = \left[\frac{(2\pi mkT)^{3/2}}{h^3}\frac{ev}{N}J(T)\right]^N \frac{e^{8.388/T^*}}{ev^*}\left(\frac{T^*}{35.01}\right)^{3/2} \times \left[1 + 10.7(v^* - 1)\exp\left(-\frac{0.0436}{T^*(v^* - 1)}\right)\right]^{N/v^*} \tag{2.29}$$

where T^* and v^* are the reduced temperature and volume, respectively, and are given by the equations

$$T^* = kT/\varepsilon, \qquad v^* = V/N\sigma^3. \tag{2.30}$$

Then, the equation of state becomes

$$\frac{pV}{NkT} = 1 - \frac{8.388}{v^*T^*} + \frac{\ln v^*}{v^*} - \frac{3}{2v^*}\ln\frac{T^*}{35.01} - \frac{1}{v^*} \times \ln\left[1 + 10.7(v^* - 1)\exp - -\left(\frac{0.0436}{T^*(v^* - 1)}\right)\right] + \frac{10.7}{T^*(v^* - 1)}\frac{0.436 + T^*(v^* - 1)}{10.7(v^* - 1) + \exp[0.436/(T^*(v^* - 1))]}. \tag{2.31}$$

The readers interested in other expressions such as excess reduced entropy (S^E/Nk) and excess reduced energy $E^E/N\varepsilon$ may read Eyring and Jhon (1969). In Fig. 5, the reduced experimental vapor pressure ($p^* = p\sigma^3/\varepsilon$) and volume

(v^*) for simple liquids such as Ne, Ar, N_2, and CH_4 are compared with the various theories and with experiment. The agreement between significant structure theory and experiment is excellent and is better than the values found for the other models plotted in Fig. 5. This is true also for other properties such as the reduced excess entropy and excess energy at the melting point (Ree *et al.*, 1965). Using the relations $(\partial P/\partial V)_T = 0$ and $(\partial^2 P/\partial V^2)_T = 0$, the critical constants can be obtained.

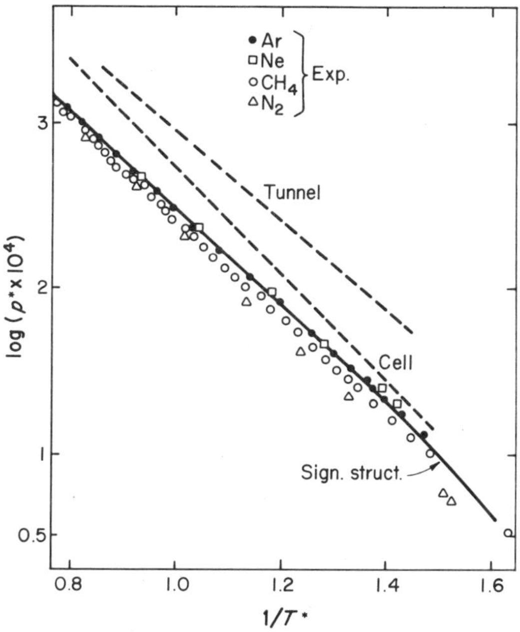

Fig. 5. Logarithm of the reduced vapor pressure p^* vs. $1/T^*$ for simple liquids. The experimental values have been reduced by using ε/k and $N\sigma^3$. The significant structure theory (——) is compared with the cell theory and tunnel theory (- - -).

In Table V, the critical properties for various theories and experiment are compared. Again, the significant structure theory is better than other theories. The cell theory does better than the tunnel theory. This may be due to the approximation of treating the motion in tunnels as strictly one dimensional. Although there are other discussions (Mayer and Careri, 1952; Dahler *et al.*, 1956; Barker, 1955, 1956, 1957, 1958), Kirkwood (1950) gives the statistical mechanical foundations for the cell theory. Therefore, it is instructive to study the correlation between models. So far, several attempts (Pierotti, 1965; Ree *et al.*, 1964a; Jhon, 1967) have been made.

Table V

Experimental and Theoretical Critical Constants[a]

	T_c^*	P_c^*	v_c^*	(PV/RT)
Mean values for Ne, Ar, N_2, CH_4	1.277	0.121	3.09	0.292
Significant structure theory	1.306	0.141	3.36	0.362
Tunnel theory	1.07	0.37	1.8	0.6
LJ & D theory[b]	1.30	0.434	1.77	0.591

[a] From Eyring and Jhon (1969, p. 43).
[b] Lennard-Jones and Devonshire.

E. Summary of the Model Concepts

Fluidized vacancies as used in significant structure theory should not be confused with the Frenkel defect in solids. The Frenkel locked-in vacancies occur in his liquid model as well as in the solid. However, the continually collapsing fluidized vacancies contribute substantially to the entropy of the system by (a) converting vibrational into translational degrees of freedom and (b) by conferring configurational degeneracy on those neighboring molecules that have sufficient kinetic energy to keep the vacancy open and to deny its use to competitors. This extra entropy makes fluidized vacancies overwhelmingly more plentiful than the usual defects of the solid. In his book Frenkel (1955) developed his idea of the liquid state. He only attempts a qualitative treatment. However, the degree of agreement with Eyring's early papers is very interesting.

In the quantitative model presented in this section, many new concepts are developed that have not appeared in earlier vacancy theories. We will summarize some of these concepts.

1. The liquid phase entirely differs from a solid phase in that the kinetic energy of molecules in the liquid has become large enough so that it can balance the potential energy tending to make molecules collapse into the holes that are present. This dynamical balance is the criteria of the melting. The entropy gained from the first appearance of holes in the solid is negligible at low concentrations of the holes but increases cooperatively as these holes become fluidized so that the Helmholtz free energy at temperatures between melting and the critical point and volumes between the solid and liquid states pass through maximum.

2. The holes or vacancies in a liquid are assumed to have the average volume of a vacancy, but are mobile, and so are called fluidized vacancies. The statistical foundations of this hypothesis have been examined recently (Choi *et al.*, 1977).

3. The fluidized vacancies in a liquid mirror the molecules in the vapor in concentration and in behavior.

4. In a mole of liquid, there are $(V - V_s)/V_s$ moles of fluidized vacancies, each of which are effective in conferring gaslike properties on V_s/V molecules. Thus, the fraction of gaslike molecules are $(V_s/V) \times [(V - V_s)/V_s] = (V - V_s)/V$, and the remainder V_s/V are solidlike molecules. V and V_s are the molar volume of liquid and solid, respectively. Very recently, Sinañoglu (1974) interpreted various types of potential between molecules in solution and liquid in terms of "significant structure" concepts.

5. Accordingly, the mean value of a liquid property X is written

$$X = X_s(V_s/V) + X_g[(V - V_s)/V]$$

where X_s and X_g are the values of the property in the solidlike and the gaslike states, respectively.

6. The vacancies, in addition to introducing gaslike properties, give the solidlike molecule extra positional degeneracy, $n_h \exp(-\varepsilon_h/RT)$ in addition to the original position. Here n_h is the number of vacancies made available to a molecule through the expenditure of the energy ε_h/N.

7. Treating the liquid at the melting point as a quasi-lattice, we find that for single spherical molecules like argon,

$$n_h = 10.7(V - V_s)/V_s, \qquad \text{and} \qquad \varepsilon_h = 0.0052E_s V_s/(V - V_s).$$

For other types of complicated molecules, the values n_h and ε_h differ in a manner to be expected.

III. Liquid Mixtures

A. Introduction

The problem of liquid mixtures is one of the most interesting areas in statistical thermodynamics. Here, we are primarily concerned with concentrated solutions where difference in molecular size is not a serious complication. Early work on the simpler molecules (Hildebrand and Scott, 1950, 1962; Guggenheim, 1952; Prigogine, 1957; Rowlinson, 1959; Eyring and Jhon, 1969; Henderson, 1974) and on high polymers (Flory, 1953; Huggins, 1962) has been treated elsewhere and so need not be repeated here. Furthermore, our discussion will be limited to binary mixtures. Then, the partition function of a mixture with N_A molecules of A species and N_B molecules of B species is given by

$$f_N = \frac{N!}{N_A!\,N_B!} \frac{(2\pi m_A kT)^{3/2}}{h^3} \frac{(2\pi m_B kT)^{3/2}}{h^3} \int e^{-\psi/kT}\, dr_A\, dr_B\, f'_N \qquad (3.1)$$

where $N = N_A + N_B$. Here f'_N is the partition function for internal degrees of freedom.

The potential energy ψ depends not only on the locations of the N molecules but also on their assignment by species to these locations. Hence, the evaluation of the partition function of a mixture is more difficult than that for the pure substance. Early, van der Waals applied the well-known equation of state to binary mixtures by simply replacing his constants a and b by the following concentration dependent averages:

$$a = x_A^2 a_A + 2x_A x_B a_{AB} + x_B^2 a_B$$
$$b = x_A b_A + x_B b_B \tag{3.2}$$

where x_A and x_B are the mole fraction of A molecules and that of B molecules, respectively. $a_{AB} = (a_A a_B)^{1/2}$, a_A and a_B are the interaction parameters for A molecules and for B molecules, b_A and b_B are a measure of the excluded volumes for both A and B species, respectively. In most models of liquid mixtures, the parameters enter as concentration weighted averages. For certain mixtures, the potential energy ψ is independent of the assignment. We call such a mixture a perfect or ideal mixture.

For an ideal binary mixture, Eq. (3.1) becomes

$$f_N = \frac{N!}{N_A!\,N_B!} f_{N_A} f_{N_B}\,. \tag{3.3}$$

Here, f_{N_A} and f_{N_B} are the partition functions of the pure species. In general, we define the thermodynamic functions of mixing by the difference between the thermodynamic properties of the mixture and those of the pure substance at the same temperature and pressure. For a perfect mixture, the Helmholtz free energy of mixing is

$$A^M = NkTx_A \ln x_A + NkTx_B \ln x_B \tag{3.4}$$

and the entropy of mixing is

$$S^M = -Nkx_A \ln x_A - Nkx_B \ln x_B \tag{3.5}$$

where the superscript M is written. One finds that A^M is negative and S^M is positive. The other properties, such as the heat of mixing and the volume of mixing, are all zero for an ideal mixture. From Eq. (3.4), one obtains

$$\mu_i - \mu_i^0 = kT \ln x_i\,. \tag{3.6}$$

Here μ_i and μ_i^0 are the chemical potential of the ith component in the mixture and that of the pure substance. Then, using the absolute activities λ

$$\lambda_i/\lambda_i^0 = x_i\,. \tag{3.7}$$

If the vapor phase in equilibrium with the mixture is assumed to be an ideal gas, then the vapor pressures p_i of the mixture are related to that of p_i^0 of the pure substance by

$$p_i/p_i^0 = x_i \tag{3.8}$$

which we call Raoult's law. In this section, we will not discuss ideal mixtures but will be concerned with the deviation from ideality.

The value of a thermodynamic function for a mixture in excess of the value for the same unmixed system will be denoted by E. The excess volume is identical with the volume change upon mixing.

B. Lattice Model; Strictly Regular Solutions

Guggenheim (1952) developed the lattice model or strictly regular solution model. Here, the term strictly regular solution is different from the term regular solution defined by Hildebrand and Scott (1950, 1962) in another context. This model assumes the following: based on the quasi-lattice picture, each molecule occupies a single lattice site. Since the lattice is assumed to be rigid, the excess volume is zero in this model. Hence, the volume is not an independent thermodynamic quantity, but is proportional to $N = N_A + N_B$.

It is assumed that the partition function for a mixture f_N can be expressed by the product of the translational partition function f_{tr} and by the internal partition function f_{int},

$$f_N = f_{tr}\, f_{int}. \tag{3.9}$$

Here,

$$f_{tr} = \left(\frac{2\pi m_A kT}{h^2}\right)^{3/2N_A} \left(\frac{2\pi m_B kT}{h^2}\right)^{3/2N_B} Q_N \tag{3.10}$$

where Q_N is the configurational partition function. The free energy of mixing is assumed to depend only on Q_N, and this term is divided into two terms as follows:

$$Q_N = Q_{latt} Q_{vib}. \tag{3.11}$$

Here, Q_{latt} is the energy of the system when each atom is at rest at a lattice site, and Q_{vib} is the vibrational energy about the sites. Since Q_{vib} is independent of the composition of the system, we only consider the effect on Q_{latt}.

We define w_{AA}, w_{AB}, and w_{BB} as the energies of each AA, AB, and BB bond, then the lattice energy is

$$E_{latt} = N_{AA} w_{AA} + N_{AB} w_{AB} + N_{BB} w_{BB} \tag{3.12}$$

where N_{AA}, N_{AB}, and N_{BB} are the number of AA, AB, and BB pairs and obey the equalities

$$2z_A = 2N_{AA} + N_{AB} \qquad 2z_B = 2N_{BB} + N_{AB}. \tag{3.13}$$

As usual, z is the coordination number of the lattice. Putting the Eq. (3.13) into Eq. (3.12), one obtains

$$E_{\mathrm{latt}} = \tfrac{1}{2}zN_{\mathrm{A}}w_{\mathrm{AA}} + \tfrac{1}{2}zN_{\mathrm{B}}w_{\mathrm{BB}} + N_{\mathrm{AB}}(w_{\mathrm{AB}} - \tfrac{1}{2}w_{\mathrm{AA}} - \tfrac{1}{2}w_{\mathrm{BB}}). \quad (3.14)$$

Introducing the definitions

$$E_{\mathrm{AA}} = \tfrac{1}{2}zN_{\mathrm{A}}w_{\mathrm{AA}}, \qquad E_{\mathrm{BB}} = \tfrac{1}{2}zN_{\mathrm{B}}w_{\mathrm{BB}} \quad (3.15)$$

$$w = w_{\mathrm{AB}} - \tfrac{1}{2}w_{\mathrm{AA}} - \tfrac{1}{2}w_{\mathrm{BB}}. \quad (3.16)$$

Then, Eq. (3.14) simplifies to

$$E_{\mathrm{latt}} = E_{\mathrm{AA}} + E_{\mathrm{BB}} + N_{\mathrm{AB}}\,w. \quad (3.17)$$

The lattice partition function becomes

$$\begin{aligned} Q_{\mathrm{latt}} &= \sum_{\langle N_{\mathrm{AB}}\rangle} g(N_{\mathrm{A}}, N_{\mathrm{B}}, N_{\mathrm{AB}}) \exp\left\{-\frac{E_{\mathrm{AA}} + E_{\mathrm{BB}} + N_{\mathrm{AB}}w}{kT}\right\} \\ &= \exp\left\{-\frac{E_{\mathrm{AA}} + E_{\mathrm{BB}}}{kT}\right\} \sum_{\langle N_{\mathrm{AB}}\rangle} g(N_{\mathrm{A}}, N_{\mathrm{B}}, N_{\mathrm{AB}}) \\ &\quad \times \exp\left\{-\frac{N_{\mathrm{AB}}w}{kT}\right\} \end{aligned} \quad (3.18)$$

where $g(N_{\mathrm{A}}, N_{\mathrm{B}}, N_{\mathrm{AB}})$ is the number of arrangements of N_{A} molecules of the first species and N_{B} molecules of the second species subject to the requirement that there be N_{AB} pairs.

Since the free energy of mixing A^{M} only depends on Q_{latt} and by the definition of A^{M}, one obtains

$$A^{\mathrm{M}} = -kT \ln \sum_{\langle N_{\mathrm{AB}}\rangle} g(N_{\mathrm{A}}, N_{\mathrm{B}}, N_{\mathrm{AB}}) \exp\left\{-\frac{N_{\mathrm{AB}}w}{kT}\right\}. \quad (3.19)$$

In the case when $w = 0$ (i.e., $w_{\mathrm{AA}} + w_{\mathrm{BB}} = 2w_{\mathrm{AB}}$), then

$$\sum_{\langle N_{\mathrm{AB}}\rangle} g(N_{\mathrm{A}}, N_{\mathrm{B}}, N_{\mathrm{AB}}) = \frac{N!}{N_{\mathrm{A}}!\,N_{\mathrm{B}}!}$$

and Eq. (3.19) becomes

$$A^{\mathrm{M}} = -kT \ln \frac{N!}{N_{\mathrm{A}}!\,N_{\mathrm{B}}!} = NkTx_{\mathrm{A}} \ln x_{\mathrm{A}} + NkTx_{\mathrm{B}} \ln x_{\mathrm{B}}. \quad (3.20)$$

Thus, for a strictly regular solution, a sufficient condition for the ideal mixture is $w = 0$. If $w \neq 0$, the sums in Eq. (3.19) must be evaluated. This mathematical problem is identical to the problem of evaluating the partition function for the well-known order–disorder problem.

In one of these approximations (i.e., Bragg–Williams approximation)

$$N_{AB} = zN\left(\frac{N_A}{N}\right)\left(\frac{N_B}{N}\right). \tag{3.21}$$

Therefore

$$A^M = NkTx_A \ln x_A + NkTx_B \ln x_B + Nzwx_A x_B. \tag{3.22}$$

Thus, the following excess functions become

$$A^E = G^E = Nzwx_A x_B \tag{3.23}$$

$$H^E = E^E = Nzwx_A x_B \tag{3.24}$$

$$S^E = 0; \tag{3.25}$$

here pV terms are neglected. Even if the pV terms are not neglected, these results are nearly the same. However, these results are inconsistent with the experimental studies, which show that the excess entropy can be positive or negative and that T times the excess entropy is of the same order as the heat of mixing. More successful models should be considered and are discussed in the following sections.

In Fig. 6, the free energy of mixing is plotted at the two temperatures, $zw/kT = 1.5$ and 2.5. It is quite natural that at the lower temperature phase separation occurs because for $w > 0$, $2w_{AB} > w_{AA} + w_{BB}$, AA and BB pairs are energetically favored over AB pairs. Applying the following critical temperature conditions $\partial^2 G^M/\partial x_A^2 = 0$, $\partial^3 G^M/\partial x_A^3 = 0$, one obtains

$$zw/kT_c = 2, \qquad x_A^c = x_B^c = \tfrac{1}{2}. \tag{3.26}$$

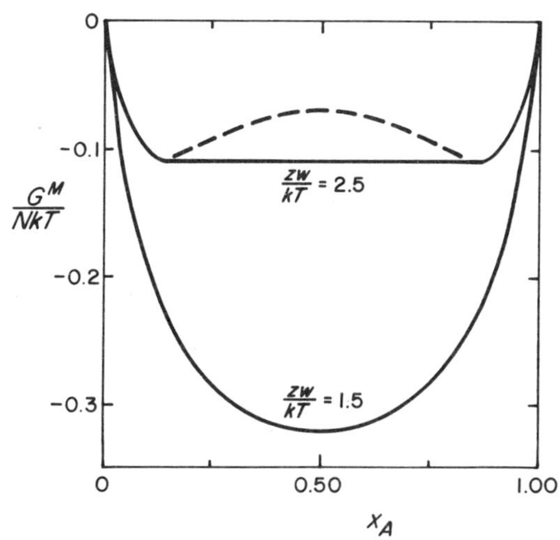

Fig. 6. Free energy of mixing according to Bragg–Williams approximation.

The results of corrections to the Bragg–William approximation and the interesting discussion on the partial vapor pressure deviation from Raoult's law are described by Guggenheim (1952).

C. Cell Model

In previous sections, one saw that the lattice model only provides qualitative information for a mixture, but fails to predict quantitatively. The free volume or cell model of Lennard-Jones and Devonshire has been applied to mixtures (Prigogine and Mathot, 1952; Salsburg and Kirkwood, 1952, 1953) with fair quantitative success. Proceeding as in Section I, configuration space is divided into N cells of equal size. This indicates that the molecules are approximately the same in size and not much different from each other. Considering the random distribution of the molecules among the cells, the configurational partition function can be written

$$f_N = \frac{N!}{N_A!\,N_B!}[ev_f(A)]^{N_A}[ev_f(B)]^{N_B} \exp\left\{-\frac{N_A\psi_A(0) + N_B\psi_B(0)}{kT}\right\} \tag{3.27}$$

where

$$v_f(A) = \int_{\text{cell}} \{\exp - [\psi_A(r) - \psi_A(0)]/kT\}4\pi r^2\, dr. \tag{3.28}$$

Similarly for $v_f(B)$, the letter A in Eq. (3.28) is replaced by the letter B. $\psi_A(r)$ and $\psi_B(r)$ are the mean potentials in a cell occupied by molecules A or B. $\psi_A(r)$ is given by

$$\begin{aligned}\int_0^\pi \{x_A u_{AA}(R) + x_B u_{AB}(R)\}\frac{1}{2} z \sin\theta\, d\theta &= \int_0^\pi 4\left[\frac{1}{R^{12}}\{x_A \sigma_{AA}\varepsilon_{AA}^{12} + x_B\varepsilon_{AB}\sigma_{AB}^{12}\}\right.\\ &\quad \left.- \frac{1}{R^6}\{x_A\varepsilon_{AA}\sigma_{AA}^6 + x_B\varepsilon_{AB}\sigma_{AB}^6\}\right]\frac{1}{2} z \sin\theta\, d\theta.\end{aligned} \tag{3.29}$$

In writing the expression for Eq. (3.29), the zx_A molecules of species A and zx_B molecules of species B are assumed to be smeared over the surface of a sphere of radius $a = (2^{1/2}V/N)^{1/3}$. If one compares Eq. (3.29) with the early Eq. (1.11), one sees that $\psi_A(r)$ is analogous to the cell potential of pure liquid having the following average parameter $\langle\varepsilon_A\rangle$ and $\langle\sigma_A\rangle$.

$$\langle\varepsilon_A\rangle = \frac{(x_A\varepsilon_{AA}\sigma_{AA}^6 + x_B\varepsilon_{AB}\sigma_{AB}^6)^2}{(x_A\varepsilon_{AA}\alpha_{AA}^{12} + x_B\varepsilon_{AB}\alpha_{AB}^{12})} \tag{3.30}$$

$$\langle\sigma_A\rangle = \left(\frac{x_A\varepsilon_{AB}\sigma_{AA}^{12} + x_B\varepsilon_{AB}\sigma_{AB}^{12}}{x_A\varepsilon_{AB}\alpha_{AA}^6 + x_B\varepsilon_{AB}\alpha_{AB}^6}\right)^{1/6}. \tag{3.31}$$

Similarly for $\psi_B(r)$, the letter A in Eqs. (3.30) and (3.31) is replaced by the letter B. Then, using the existing tables for calculating properties for pure liquid, the mixture properties can be evaluated. On the other hand, one can simplify the expression somewhat by assuming that the cells are equivalent on the average, and that a single free volume may be used. We call this a single-fluid model, in contrast to the two-fluid model we have already considered. The predictions of the two models are not greatly different.

The configuration partition function is now

$$f_N = \frac{N!}{N_A!\,N_B!}\{ev_f\}^N \exp\frac{N\psi(0)}{kT} \tag{3.32}$$

where

$$v_f = \int_{\text{cell}} \exp\left\{-\frac{\psi(r)-\psi(0)}{kT}\right\} 4\pi r^2\, dr$$

and

$$\begin{aligned}\psi(r) &= x_A\psi_A(r) + x_B\psi_B(r)\\ &= \int_0^\pi \{x_A^2 u_{AA}(R) + 2x_A x_B u_{AB}(R) + x_B^2 u_{BB}(R)\}\tfrac{1}{2}z \sin\theta\, d\theta.\end{aligned} \tag{3.33}$$

This expression is analogous to the cell potential of a pure liquid having the following average parameters:

$$\langle\varepsilon\rangle = \frac{(x_A^2\varepsilon_{AA}\sigma_{AA}^6 + 2x_A x_B\varepsilon_{AB}\sigma_{AB}^6 + x_B^2\varepsilon_{BB}\sigma_{BB}^6)^2}{(x_A^2\varepsilon_{AA}\sigma_{AA}^{12} + 2x_A x_B\varepsilon_{AB}\sigma_{AB}^{12} + x_B^2\varepsilon_{BB}\sigma_{BB}^{12})} \tag{3.34}$$

$$\langle\sigma\rangle = \left(\frac{x_A^2\varepsilon_{AA}\sigma_{AA}^{12} + 2x_A x_B\varepsilon_{AB}\sigma_{AB}^{12} + x_B^2\varepsilon_{BB}\sigma_{BB}^{12}}{x_A^2\varepsilon_{AA}\sigma_{AA}^6 + 2x_A x_B\varepsilon_{AB}\sigma_{AB}^6 + x_B^2\varepsilon_{BB}\sigma_{BB}^6}\right)^{1/6}. \tag{3.35}$$

The model predicts much better agreement with experiment than the lattice model (or strictly regular solution) does. Particularly, the signs of the excess thermodynamic functions are predicted well. However, the magnitude of these quantities usually shows considerable error. In Table VI in the next section, some of the calculated excess properties for neopentane–CCl_4 are compared with experiment.

The main defects of the cell model can be overcome by using more satisfactory models of the liquid state and utilizing more realistic and appropriate potential functions. Among such theories, the extension of the tunnel theory of liquids to a mixture system is noteworthy. The tunnel theory is often more satisfactory than the cell model. Malijevský (1971) has developed

Barker's tunnel model of the liquid for many binary mixtures. The calculations were carried out using the Lennard-Jones (6-12) potential, and the parameters relating unlike molecules were calculated from Lorentz's and Berthelot's combination rules. They found that the agreement between theory and experiment for the excess function is somewhat better than that for the cell model used by Prigogine. Another effort utilizes improved expressions for the average potential parameters. This concept is the basis of Prigogine's average potential model.

D. Average Potential Model

A less exciting, but still interesting, procedure is introduced by Prigogine *et al.* (1956) using the experimental corresponding state thermodynamic functions. For convenience, we call it the average potential model. According to the theorem of corresponding states for a pure liquid, the configurational partition function Q_N is given by

$$Q_N = [\sigma^3 q(T^*, v^*)]^N \tag{3.36}$$

where $q(T^*, v^*)$ is a universal function of $T^* = kT/\varepsilon$ and $v^* = V/N\sigma^3$. In the crude approximation

$$Q_N = \frac{N!}{N_A!\,N_B!}[\langle\sigma\rangle^3 q(\langle T^*\rangle, \langle V^*\rangle)]^N \tag{3.37}$$

where $\langle T^*\rangle = kT/\langle\varepsilon\rangle$, $\langle v^*\rangle = V/N\langle\sigma\rangle^3$, and $\langle\varepsilon\rangle$ and $\langle\sigma\rangle$ are given in Eqs. (3.34) and (3.35). This is visualized as a one-fluid model.

In Table VI, the excess functions evaluated by the cell model using the average potential model for a neopentane–CCl_4 mixture are compared with experiment (Mathot and Desmyter, 1953; Englert-Chowles, 1955). Both models predict the correct sign, but the average potential model is more

Table VI

Calculated and Measured Excess Properties of a Neopentane–CCl_4 Mixture at 0° C ($x = 0.5$)

	G^E (cal mole^{-1})	H^E (cal mole^{-1})	TS^E (cal mole^{-1})	V^E (cm^3 mole)
Cell model		16	−98	−3
Average potential model	80	71	−13	−2.2
Experiment	76	75	−125	−0.5

satisfactory. Refined versions of the average potential model in which non-randomness and local inhomogeneities of structure are taken into account have been formulated (Prigogine, 1957). This is equivalent to a two-fluid model. The difference between the two calculated results is slight. More sophisticated techniques are treated in detail by Prigogine (1957).

Recently, utilizing some concepts related to those in the "quasi-chemical" treatment (Guggenheim, 1952), as further developed by Prigogine (1957) and refining the treatment of Hildebrand and Scott (1950, 1962), Huggins (1970) has developed the theory for the intermolecular energies in a mixture. Although only a limited number of properties such as heat of mixing are evaluated, the treatment gives better agreement with experiment than other models. According to him, the intermolecular energies in a mixture are related to the intermolecular contact surface areas of the segments (σ) of which the molecules are composed, to the contact energies (ε) per unit area, and to an equilibrium constant (K) relating different contacting surface areas. In his treatment, each substance is assumed to be monatomic containing only one type of contacting group. The following expression for the intermolecular molar energy E is used:

$$E = \frac{\sigma_\alpha \varepsilon_{\alpha\alpha}}{2} + \frac{\sigma_\beta \varepsilon_{\beta\beta}}{2} - \frac{(\sigma_\alpha + \sigma_\beta)\,\Delta\varepsilon K}{4(1-K)}[1 - (1+y)^{1/2}] \tag{3.38}$$

where

$$K = \alpha_{\alpha\beta}^2 / 4\alpha_{\alpha\alpha}\alpha_{\beta\beta} \tag{3.39}$$

$$y = \frac{4\alpha_\alpha \alpha_\beta (1-K)}{(\alpha_\alpha + \alpha_\beta)^2 K} \tag{3.40}$$

$$\Delta\varepsilon = 2\varepsilon_{\alpha\beta} - \varepsilon_{\alpha\alpha} - \varepsilon_{\beta\beta}\,. \tag{3.41}$$

Here, two types of segments in a mixture are designated by subscripts α and β, and the types of segment–segment contacts by subscripts $\alpha\alpha$, $\alpha\beta$, and $\beta\beta$.

E. Significant Liquid Structure Model

A series of papers have used significant structure theory to treat the following binary mixtures successfully: $CCl_4 + C_6H_{12}$ (Liang *et al.*, 1964), $C_6H_{12} + C_6H_6$ (Ma *et al.*, 1965), $A + N_2$ (Miner and Eyring, 1965), $A + O_2$ (Miner and Eyring, 1965), $O_2 + N_2$ (Miner and Eyring, 1965), $C_6H_6 + C_2H_4O_2$ (Ahn, Pack, and Chang, 1965), alkali halide mixtures (Vilen and Misdolea, 1966), and cryogenic liquid mixture (only surface tension) (Kim *et al.*, 1968; Kim, Eyring, and Lee, 1969). According to the significant liquid structure model, there are three significant structures in a liquid: (1) molecules with solidlike degrees of freedom, (2) positional degeneracy in the solidlike structure, and (3) molecules with gaslike degrees of freedom. The theory can be extended directly to binary mixtures by considering the concentration dependence of the parameters. Also, the following assumptions are made: (a) nonrandom mixing is negligible in the case considered; (b) the same characteristic temperatures of vibration θ are retained for mixtures that were used for the pure substances; (c) molecules of both components continue to possess their gaslike translational degree of freedom; (d) the degeneracy term has the same form as for a pure liquid; (e) the parameters E_s, V_s, n, and a may be taken as a suitable average of the parameters for the pure components. With these assumptions, the partition functions for a mixture takes the following form:

$$f_{\text{mixture}} = \frac{(N_1 + N_2)!}{N_1!\,N_2!}[f_{s1}^{N_1} f_{s2}^{N_2}]^{V_s/V} \times [f_{\text{deg}} \exp(E_s/RT)]^{(N_1+N_2)V_s/V}[f_{g1}^{N_1} f_{g2}^{N_2}]^{(V-V_s)/V} \tag{3.42}$$

where

$$f_{s_1} = \frac{1}{[1 - \exp(-\theta_T)]^3} f_{\text{rot }1}^{(s)} f_{\text{vib }1}\,. \tag{3.43}$$

Here, we assume that the solidlike molecule rotates

$$f_{g_1} = \frac{(2\pi mkT)^{3/2}}{h^3} f_{\text{rot }1}^{(g)} f_{\text{vib }1}\left(\frac{eV}{N_1 + N_2}\right) \tag{3.44}$$

with similar equations for component 2, and

$$f_{\text{deg}} = 1 + n\left(\frac{V - V_s}{V_s}\right)\exp\left(\frac{-aE_s V_s}{(V - V_s)RT}\right) \tag{3.45}$$

$$E_s = x_1^2 E_{s_1} + x_2^2 E_{s_2} + 2(1 + \delta E)x_1 x_2 (E_{s_1} E_{s_2})^{1/2} \tag{3.46}$$

$$V_s = x_1 V_{s_1} + x_2 V_{s_2} + x_1 x_2\, \delta_V (V_{s_1} V_{s_2})^{1/2} \tag{3.47}$$

$$n = x_1 n_1 + x_2 n_2 \tag{3.48}$$

$$a = x_1 a_1 + x_2 a_2\,. \tag{3.49}$$

The quantities δ_E and δ_V are the only parameters in the mixture partition function that were not evaluated from the pure liquids. The correction parameters, δ_E and δ_V in the cross terms are introduced as higher order effects of the concentration. Otherwise the notation has been defined previously. Since the terms involving the deltas are generally very small, reasonable results are obtained when they are assumed to be zero.

At constant temperature, the equilibrium condition satisfies the following sets of relationships:

$$p^L = p^G; \qquad -(\partial A/\partial V)^L_{T,N_1,N_2} = -(\partial A/\partial V)^G_{T,N_1,N_2} \tag{3.50}$$

$$\mu_1^L = \mu_1^G; \qquad (\partial A/\partial N_1)^L_{T,V,N_2} = (\partial A/\partial N_1)^G_{T,V,N_2} \tag{3.51}$$

$$\mu_2^L = \mu_2^G; \qquad (\partial A/\partial N_2)^L_{T,V,N_1} = (\partial A/\partial N_2)^G_{T,V,N_1} \tag{3.52}$$

where μ_i is the chemical potential of the ith component. To evaluate the thermodynamic properties and the excess properties of a mixture, one first must find two points on the free energy surface which satisfy the three conditions in Eqs. (3.50)–(3.52). Some of the results for the excess quantities are shown in Fig. 7 and Table VII.

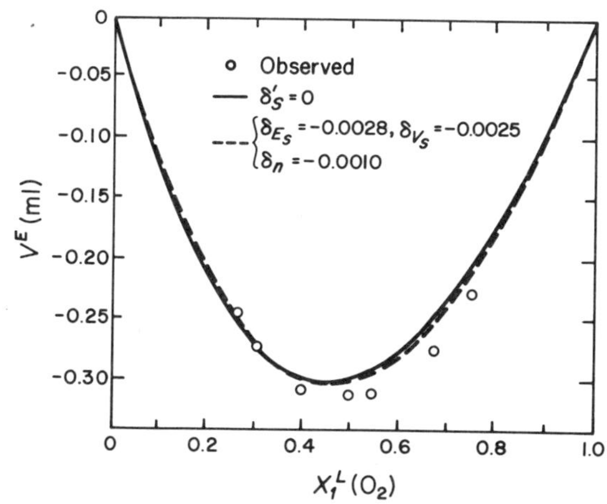

Fig. 7. Excess volumes of the $O_2 + N_2$ system at 83.82°K and constant pressure.

Agreement between theory and experiment is quite satisfactory. To be a successful model for a mixture, one must explain the phase separation problem. Very recently, Lee *et al.* (1977) extended the model to predict the composition–temperature dependence of the O_2–O_3 systems with success. In their calculations, the following criteria of the phase separation is used:

$$\left(\frac{\partial^2 G_{\text{mixing}}}{\partial x^2}\right)_{T,p} = \left(\frac{\partial^2 G_{\text{mixture}}}{\partial x^2}\right)_{T,p} < 0. \tag{3.53}$$

Table VII

Excess Properties of the Ar + N_2 System at 83.82°K and Constant Pressure[a]

X_1^1 (Ar)	G^E (cal/mole)			H^E (cal/mole)			TS^E (cal/mole)		C_p^E (cal/mole°K)	
	Calc.		Obs.	Calc.		Obs.	Calc.		Calc.	Obs.
0.1156	2.00	3.23	—	3.00	4.73	5.6	1.02	1.52	−0.132	−0.125
0.2243	3.45	5.55	—	5.29	8.20	9.7	1.86	2.66	−0.242	−0.232
0.3482	4.53	7.30	—	7.21	10.99	11.6	2.68	3.70	−0.346	−0.334
0.4315	4.94	7.93	—	8.06	12.13	12.2	3.12	4.20	−0.400	−0.389
0.5000	5.05	8.12	8.2	8.44	12.59	12.1	3.39	4.47	−0.432	−0.422
0.5420	5.03	8.08	—	8.54	12.66	12.0	3.50	4.56	−0.445	−0.437
0.6110	4.84	7.77	—	8.44	12.38	11.4	3.59	4.59	−0.454	−0.448
0.7229	4.10	6.59	—	7.52	10.84	9.8	3.39	4.23	−0.429	−0.426
0.8151	3.10	4.98	—	5.95	8.46	7.5	2.83	3.45	−0.358	−0.359
0.9387	1.10	1.87	—	2.46	3.43	2.7	1.28	1.52	−0.161	−0.163

Column 1: $\delta_v = 0$, $\delta_E = 0$; Column 2: $\delta_E = -0.0043$, $\delta_V = -0.009$; Column 3; observed

[a] From Eyring and Jhon (1969, p. 72).

The results are shown in Fig. 8. One sees that the model of free rotation of solidlike molecules of O_2 and O_3 is an approximation, and there would be improvement if the more appropriate hindered rotation for the solidlike molecules were taken into account. A hindered rotation function f_{HR}, having the following form has been used elsewhere.

$$f_{HR} = f_{vib} + (f_{FR} - f_{vib}) \exp\left[-\frac{BV_s}{RT(V - V_s)}\right]. \tag{3.54}$$

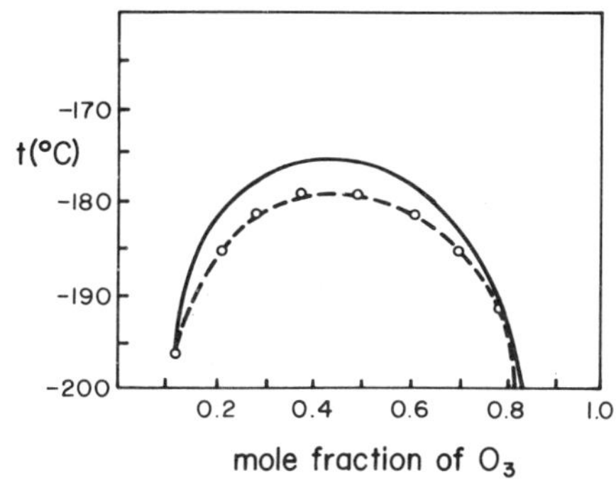

Fig. 8. Phase diagram of O_3 (temperature vs. mole fraction).

Here f_{FR} is the free rotational partition function, and f_{vib} is the restricted vibrational partition function which has sometimes been taken as unity. V_{s0} is the solid volume at which free rotation begins, and the constant B is the height of the potential barrier at which rotation begins.

IV. Thermodynamic and Transport Properties

A. Introduction

A successful liquid model should agree with the Monte Carlo (Wood, 1968) and molecular dynamics (Wainwright and Alder, 1958) calculations for all thermodynamic and transport properties for the limited number of cases where comparison is possible. However, a realistic model should also explain all thermodynamic and transport properties of real liquids. In this respect, the most successful model is the significant liquid structure model (Eyring and Jhon, 1969). Very few applications for real liquid are available using the other models.

In subsequent sections we discuss the application of models in calculating the properties of inorganic and organic liquids, quantum liquids, fused salts and liquid metals, liquid crystals and high polymers. Finally, we discuss the transport properties of liquid such as viscosity and thermal conductivity. In actual thermodynamic calculations a noteworthy merit of the significant liquid structure model is its usefulness in calculating values for the liquid density and the vapor pressure, an aspect in which the other models in general fail.

If the Helmholtz free energy A is plotted as a function of volume at constant temperature, as shown in Fig. 9, and if the common tangent to the

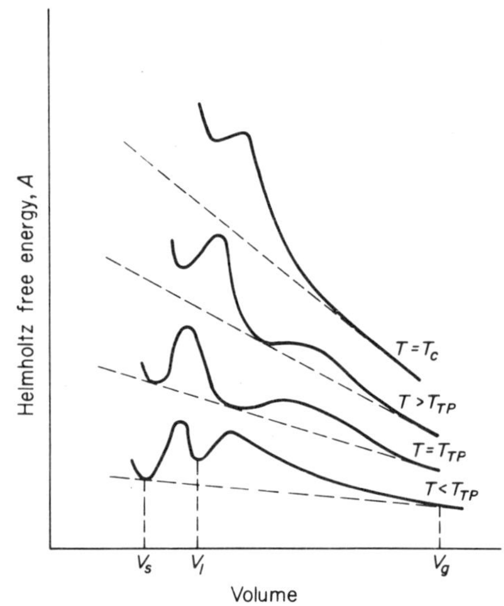

Fig. 9. Isotherms for the Helmholtz free energy A vs. volume. The scales are greatly distorted. T_c is the critical temperature and T_{TP} is the triple point temperature.

points corresponding to the liquid and vapor phase is determined by calculation, the vapor pressure is given by the slope of the common tangent and the abscissas of the two points of tangency indicate the respective volumes of the liquid and vapor. Other thermodynamic properties are obtained using the appropriate equations such as

$$S = -(\partial A/\partial T)_V, \qquad E = -T^2(\partial(A/T)/\partial T)_V, \qquad C_v = T(\partial S/\partial T)_V \quad \text{etc.} \tag{4.1}$$

for all liquids.

B. Inorganic and Organic Liquids

The significant liquid structure model has been applied to many inorganic and organic liquids (Eyring and Jhon, 1969) ranging from simple monoatomic liquids such as argon to complicated ones such as pentaborane and from simple hydrocarbons such as methane to more complicated molecules such as the xylenes (Jhon *et al.*, 1965) and halobenzene (Wu *et al.*, 1972). In each individual case, the partition function for these liquids were addressed to the specific properties of the liquid. For example, the rotational term with a triplet ground state is included in both the gaslike and solidlike parts of the partition function for oxygen (Kim *et al.*, 1967), while a six-degree Einstein oscillator term appears in the solidlike portion while the rotational term appears in the gaslike partition function for diborane (Jhon *et al.*, 1967b) since this molecule fails to rotate freely in the solid state. The criteria for free rotation or nearly free rotation in the solidlike degrees of freedom are sometimes difficult to apply. In general, it is reasonable to assume a hindered rotational degree of freedom for nonspherical molecules near the melting point. McLaughlin and Eyring (1966) introduced an approximate hindered rotational partition function f_{HR} into f_s [see Eq. (3.54)]. They used as the f_s for N_2 and Cl_2

$$f_s = \frac{\exp(E_s/RT)}{(1 - e^{-\theta/T})^3} (f_{HR})^2 \tag{4.2}$$

and for C_6H_6, CH_4, CCl_4, and CH_3Cl,

$$f_s = \frac{\exp(E_s/RT)}{(1 - e^{-\theta/T})^3} (f_{HR})^3. \tag{4.3}$$

C. Quantum Liquids; Liquid Hydrogen

Although the cell model is crude, it is one of the few liquid models that is simple enough to yield numerical results. Hence, there have been several attempts to introduce quantum effects into the cell model. It has been found that the reduced properties of hydrogen are not satisfactory using the classical cell model of Lennard-Jones and Devonshire. The deviations have been explained as quantum effects (Hamann, 1952, 1954; David and Hamann, 1953). De Boer and Lunbeck (1948) have expanded the Slater sum in terms of the square of the quantum parameter which value is greater than unity and fails to apply to hydrogen. Hamann (1952, 1954) and David and Hamann (1953) replaced the actual potential by a square potential well, replaced the spherical cell by a cubic cell, and the partition function is approximated by a series of integrals. The method is surprisingly reliable. Levelt and Hurst (1960) have solved numerically Schrödinger's equation for the actual Lennard-Jones and Devonshire potential. However, the numerical results are not readily obtained. It is therefore useful to calculate the energy levels of a molecule in a cell by an approximate procedure yielding accurate results more quickly than the method of Levelt and Hurst. Henderson and Reed (1964) applied the Wenzel, Kramers, and Brillouin (WKB) method to obtain the energy levels of liquid hydrogen taking into account the first three shells of neighbors.

The properties obtained by the above models give very similar results. Significant structure theory has also been applied to quantum liquids such as liquid hydrogen, deuterium, and hydrogen deuteride (Henderson *et al.*, 1962). The partition functions involve different forms of statistics, and changes in the concentrations of ortho- and para-hydrogen also cause slight changes in the thermodynamic properties. Because these liquids exist at extremely low temperatures, a Debye partition function was used for the solidlike molecules. At the temperatures of interest, Bose–Einstein statistics are required for hydrogen and deuterium and Fermi–Dirac statistics for hydrogen deuteride. The following partition function is obtained:

$$\ln f = N\frac{V_s}{V}\left\{\frac{E_p}{RT} - \frac{9}{8}\left(\frac{\theta_D}{T}\right) - 9\left(\frac{T}{\theta_p}\right)^3 \int_0^{\theta_{D/T}} u^2 \ln(1 - e^{-u})\, du + \ln\left(1 + Z\frac{V - V_s}{V}e^{-\alpha}\right)\right\} + N\frac{V - V_s}{V}\left(1 - \ln y \pm \frac{y}{2^{5/2}}\right) + N \ln f_r \tag{4.4}$$

where the top sign (+) applies to the Bose–Einstein case and the bottom sign (−) in the Fermi–Dirac case; here $\alpha = E_p V/[RT(V - V_s)]$, and E_p is the potential energy of the lattice, θ_D is the Debye temperature of the solidlike lattice, $y = [Nh^3/V(2\pi mkT)^{3/2}]$, and f_r is the sum of the rotational partition functions for the ortho and para species. Table VIII lists some calculated and observed properties at the melting, boiling, and critical points.

D. Molten Metals and Fused Salts

In spite of their practical and theoretical importance, few theoretical models have been developed for the fused salts. Carlson *et al.* (1960a) applied the significant liquid structure model to fused salts with success. Blomgren (1960) also applied an extended significant structure model to

Table VIII

CALCULATED AND OBSERVED THERMODYNAMIC PROPERTIES OF LIQUID HYDROGEN[a]

	p-H_2	n-H_2	H-D	o-D_2	n-D_2	
T_m (°K)	(13.84)	(13.94)	(16.60)	(18.63)	(18.73)	calc
	13.84	13.94	16.60	18.63	18.73	obs
P_m (atm)	0.07388	0.07589	0.1236	0.1706	0.1724	calc
	0.06942	0.07085	0.1221	0.1678	0.1692	obs
V_m (cm^3 mole^{-1})	26.213	26.093	24.491	23.262	23.155	calc
	26.176	26.108	24.487	—	23.162	obs
ΔS_m (cal	1.932	1.936	2.048	2.198	2.210	calc
mole^{-1} deg^{-1})	2.028	—	—	2.526	—	obs
T_b (°K)	20.58	20.70	22.29	23.65	23.75	calc
	20.261	20.365	22.14	23.59	23.67	obs
V_b (cm^3 mole^{-1})	28.829	28.692	26.525	24.955	24.830	calc
	28.482	28.393	—	—	—	obs
ΔS_b (cal	10.553	10.564	11.868	12.741	12.737	calc
mole^{-1} deg^{-1})	10.602	—	—	12.459	—	obs
T_c (°K)	35.9	36.2	37.6	39.4	39.7	calc
	32.994	33.24	35.908	38.262	38.24	obs
P_c (atm)	13.6	13.8	15.5	17.1	17.3	calc
	12.770	12.797	14.645	16.282	16.421	obs
V_c (cm^3 mole^{-1})	77.7	77.3	71.5	68.3	68.0	calc
	65.5	—	62.8	60.3	—	obs

[a] From Eyring and Jhon (1969, p. 65).

fused salts. A treatment of slightly ionized salts by Jhon *et al.* (1968) and an extension of Carlson's work by Lu *et al.* (1968) are of interest. Another successful application of the cell model by McQuarrie (1962) should be noted. He applied the cell model in a straightforward manner to systems of fused salts. Because of the long range of the Coulombic potential, the calculations are extended to include all neighboring shells. This problem is resolved by the natural occurrence of the Madelung constant. The result shows that the Coulomb potential does not influence the motion of an ion directly, so that a molten salt may be approximated as a fluid of hard spheres in certain cases.

Several thermodynamic properties including critical volumes are obtained for molten alkali halides. The following points arose in treating the molten alkali halides using the significant liquid structure model (Carlson *et al.*, 1960a). It is known that the volume expansion upon melting is approximately twice that of argon. This is expected since the entropy of melting comes from the randomness introduced by the excess volume, and only half the vacancies are available to positive and half to negative ions. Thus, twice the percentage expansion is required to give the same entropy increase for each kind of ion. Since only half the excess volume ($V - V_s$) provides vacancies for each kind of ion, we expect the proportionality constant n in the degeneracy term in the partition function to be only about half as large for molten salts as it is for argon. In constructing the partition function, the solidlike portion of the partition function is raised to the power $2N(V_s/V)$, since a mole of alkali halides contains $2N$ particles. E_s/N is the sublimation energy per alkali halide molecule, and must be divided by two to get the required E_s per ion. Due to the long range nature of the Coulombic interionic potential, the solid E_s is multiplied by the factor $(V/V_s)^{1/3}$. In the original treatment, the gaslike structure is assumed to be only diatomic alkali halide molecules. However, the results have been improved by considering the equilibrium between monomer and dimer in the vapor phase (Lu *et al.*, 1968).

The liquid metals are generally considered to be monatomic simple liquids. The significant structure model was first applied to liquid metals by Carlson *et al.* (1960b). Later, Vilen and Misdolea (1968) extended these calculations using Carlson's partition function to evaluate other thermodynamic properties such as C_p and C_v. In liquid metals the positive ions rather than the atoms move into fluidized vacancies as though they were dissolved in a sea of conducting electrons. Since the metal ions are about $\frac{1}{3}$ to $\frac{1}{4}$ as large as the metal atoms, this explains why the same entropy of melting is obtained with a correspondingly smaller expansion. The entropy of melting in metals is about the same as in the melting of normal liquids. A successful hole theory for liquid metals must use this factor of small hole size in explaining the degeneracy introduced upon melting. For the construction of the partition function of a liquid metal, an Einstein solid is assumed for the solidlike degree of freedom, and both monatomic and diatomic contributions are included in the gaslike degrees of freedom since a significant amount of the dimer exists in the vapor.

To improve the earlier calculation of liquid metals, Hsu *et al.* (1972) introduced the solid volume and temperature dependence of the E_s and θ in the partition function. The results are quite satisfactory. In actual calculations, the following equations were used:

$$E_s = E_{s0}[2(V_{s0}/V_s)^q - (V_{s0}/V_s)^{2q}] \tag{4.5}$$

$$\theta = \theta_0(V_{s0}/V_s)^\gamma \tag{4.6}$$

$$V_s = V_{s0}(1 + \alpha_s T). \tag{4.7}$$

The value of q was taken from the best solid state data, γ is the Gruneisen constant, and α_s is the expansion coefficient of solidlike molecules. The parameters E_{s0}, V_{s0}, and θ_0, which are the quantities at absolute zero temperature, were fixed to give the best liquid state parameter of E_s, θ, and V_s at the melting point.

E. Polymers and Liquid Crystals

The simplified version of the significant liquid model is still useful for explaining the properties of complicated molecules such as polymers and liquid crystals. Stewart and von Frankenberg (1967, 1968) applied the model to amorphous polyethylene to evaluate some thermodynamic properties. However, they used an empirical formula to compute volume, did not report solid properties, and did not consider transport properties. These facts led to the introduction of a modified model by Ma *et al.* (1974) that takes into account all properties of polyethylene. Polyethylene is a linear polymer having the formula $(CH_2)_{n-2}(CH_3)_2$. The end effect of the CH_3 group of the molecule is neglected in comparison with the periodic long CH_2 structure. It is assumed that, above the softening point, holes appear in the solid lattice of amorphous polyethylene. The kinetic segments, each containing some 20 to 25 CH_2 units can move into these vacancies. There are $n_1 = M/m_1$ kinetic segments in a molecule, where M is the molecular weight of the polymer, and m_1 is the mass of the kinetic segment. Furthermore, a polymer molecule is obliged to slalom around $n_2 - 1$ tangle points where $n_2 = M/m_2$ and m_2 is the average mass between tangle points, in order that the molecules center of gravity can retreat or advance along the shear plane. Hence, the term $(M/m_1)^{4/3}(M/m_2)^2$ is added as a factor in the normal liquid partition function. The positional degeneracy term f_{deg} is present due to the translational motion of kinetic segments, with each segment containing n CH_2 units, and also due to the libration of the individual CH_2 units.

The solidlike partition function, f_{solid}, is constructed by making use of the complete frequency distribution for the 100% crystalline solid which has been calculated by Wunderlich (1962) except that in addition the terms that take into account the hydrogen vibrations are included. A similar form to that for the solidlike molecule is used for gaslike molecules except that the three degrees of freedom, the accoustical vibrations, are replaced by three translational degrees of freedom. Then, for one mole of amorphous polyethylene, the total partition function f can be summarized as follows;

$$
\begin{aligned}
f = {} & [f_{\text{solid}}\, f_{\text{deg}}]^{N(M/m_1)(V_s/V)} \\
& \times \left[[f_g]^{N(M/m_1)[(V-V_s)/V]} \Big/ \left(N\frac{M}{m_1}\frac{V-V_s}{V} \right)! \right] \\
& \times \left[\left(\frac{M}{m_1}\right)^{4/3} \left(\frac{M}{m_2}\right)^{2} \right]^{N}.
\end{aligned}
\tag{4.8}
$$

The partition function for crystalline polyethylene can be derived by dropping the last term and putting $V = V_s$ in Eq. (4.8). Good results are obtained for thermodynamic properties as well as transport properties. Figure 10 shows calculated volumes of amorphous and crystalline polyethylene along with the experimental curves. Recently, the significant liquid structure model (Ma and Eyring, 1975; Do *et al.*, 1976) has again been used to explain the

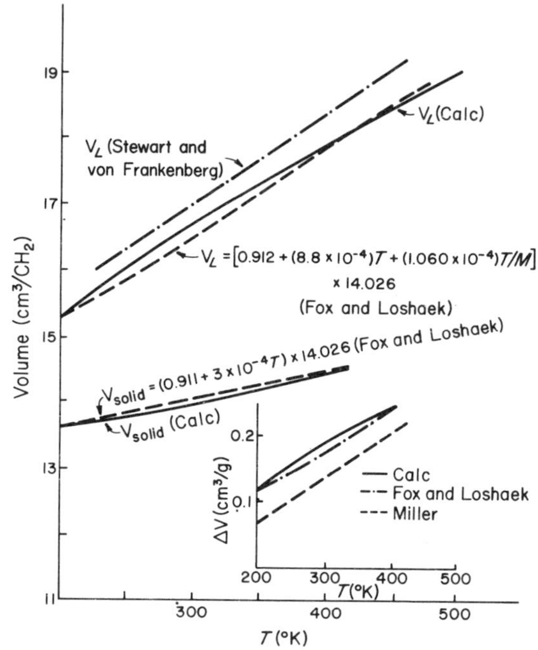

Fig. 10. Volume of amorphous and crystalline polyethylene.

properties of the nematic *p*-azoxyanisole. The two procedures have many features in common. In the latter treatment the order–disorder partition function is introduced in addition to the normal partition function (Do *et al.*, 1976).

F. Viscosity and Diffusion

Most reasonably successful models of liquid viscosity have been based on the assumption of a quasi-crystalline liquid structure. The viscosity of liquids at temperatures above their melting points are quite well described by an Arrhenius-type equation,

$$\ln \eta = A + E/RT \tag{4.9}$$

where A is a constant, R is the gas constant, T is the absolute temperature, and E is the activation energy for viscous flow. The same Arrhenius form is obtained theoretically using Eyring's rate theory (Glasstone *et al.*, 1941). However, this approach must be modified for liquids of high viscosity or for highly supercooled liquids. This fact stimulated the currently active free volume approach to liquid viscosities.

Doolittle (1951, 1952) and Williams *et al.* (1975) have empirically shown that the viscosity can be related to the free volume of the liquid by the relation

$$\eta = A \exp(B/f) \tag{4.10}$$

where f is the fractional free volume and A and B are constants, the latter being about unity. Later, Cohen and Turnbull (1959) developed this expression on a theoretical basis. However, a free volume theory of this kind fails to explain the temperature and pressure dependence of many liquids. To overcome the defects of Eqs. (4.9) and (4.10), Macedo and Litovitz (1965) proposed a hybrid form of both equations. However, their equation is not related to other properties of the liquid state.

Using the significant liquid structure model (Eyring and Jhon, 1969), Eyring and Ree (1961) applied the absolute reaction rate theory to various transport phenomena. Their equation includes both temperature and volume dependence of the viscosity. Furthermore, the model was extended to the supercooled region and to the prediction of a glass transition temperature (Faerber *et al.*, 1970; Bahng *et al.*, 1971). We will discuss such models in more detail.

1. The Free Volume Model of Cohen and Turnbull

Assuming a distribution of hole sizes exists and the probability of jumping is determined only by the chance of finding an adjacent local free volume of sufficient size to jump into, Cohen and Turnbull derived their Eq. (4.10). They assumed that the liquid consisted of hard spheres with only repulsive forces, and each molecule is confined in a cage surrounded by its neighbors. For diffusive motion to occur, there must appear a fluctuation in the local density that provides a hole large enough to allow a significant displacement of the molecule within the cage. They associate a local free volume with each molecule and assume a distribution of local free volumes. In this way, the molecule can be characterized by their associated free volume v_f.

They also obtained an expression for the excess entropy due to the distribution of v_f and minimized the excess entropy holding the number of molecules and total free volume constant. In this way the hole size distribution $p(v)$ is given as

$$p(v) = (\gamma/v_f) \exp(-\gamma v/v_f) \tag{4.11}$$

where $p(v)$ is the probability of finding the free volume v nearby, v_f is the average free volume per molecule, and γ is the numerical factor for the correction of the overlapping free volume which lies between $\frac{1}{2}$ and 1. Assuming that some minimum local free volume v^* is necessary for a jump to occur, one can calculate the probability of finding v^* and thus the jump probability. The quantity P_v is given by

$$P_v = \int_{v^*}^{\infty} p(v)\, dv = \exp\left(\frac{-\gamma v^*}{v_f}\right). \tag{4.12}$$

Since the quantity v^* should be close to v_0, the close-packed molecular volume, the hard sphere viscosity equation takes the form

$$\eta = A/P_v = A \exp(\gamma v_0/v_f). \tag{4.13}$$

The model fails to explain the temperature and pressure dependence of many liquids. Equation (4.13) predicts that $(\partial \ln \eta/\partial T) = 0$ or positive, which is not observed in many cases. However, it has been pointed out that the free volume has both a pressure and a temperature dependence (Miller, 1964; Matheson, 1966). It is interesting that Cohen and Turnbull have improved their free volume model by dividing displacements into gaslike and solidlike categories in accord with the significant liquid structure model (Turnbull and Cohen, 1970).

2. The Significant Liquid Structure Model

The significant structure theory of liquids has implicit in it a general theory of transport properties (Eyring and Jhon, 1969). According to the model, the fraction V_s/V of the shear plane is covered by solidlike molecules, and the rest $(V - V_s)/V$ is covered by gaslike molecules. Then, the viscosity η is given by

$$\eta = (V_s/V)\eta_s + [(V - V_s)/V]\eta_g. \tag{4.14}$$

The gaslike viscosity η_g is derived from the kinetic theory of gases (Moore, 1962) and is

$$\eta_g = (\tfrac{2}{3}d^2)(mkT/\pi^3)^{1/2} \tag{4.15}$$

where d is the diameter of the molecules. Using Eyring's early viscosity theory (Eyring, 1936), and assuming Newtonian flow in the viscous flow-rate processes, Ree *et al.* (1962) derived the following equation for the viscosity of solidlike molecules.

$$\eta_s = \frac{Nh}{Z\kappa'}\left(\frac{6}{\sqrt{2}}\right)\frac{V}{V_s}\frac{1}{(V - V_s)}\frac{1}{(1 - e^{-\theta/T})}\exp\left(\frac{a'E_s V_s}{RT(V - V_s)}\right). \tag{4.16}$$

Here κ' and a' are the transmission coefficient and proportionality constant, and the other symbols have already been defined. For the hard sphere system, an early paper gives a derivation (Ree *et al.*, 1964b). Equation (4.14) is self-consistent since if V approaches V_s, then η becomes infinite and as V grows compared to V_g, η is reduced to η_g. The effect of pressure was also taken into account in the theory of Ree *et al.* (1962) and Jhon *et al.* (1969).

By using the theory of Eyring *et al.* (1958b), relating high polymer viscosity to that of small molecules, Eq. (4.14) has been extended into the high polymer range. The model for the viscosity of a pure liquid is readily adapted to mixtures. The binary liquid partition function can be used to develop the viscosity equations for mixtures (Bahng *et al.*, 1970). The viscosity formula being considered is

$$\eta = \frac{hN}{Z\kappa'}\frac{6}{2^{1/2}}\frac{1}{(1-e^{-\theta_1/T})^{X_1}}\frac{1}{(1-e^{\theta_2/T})^{X_2}}$$
$$\times \exp\frac{bE_s V_s}{RT(V-V_s)}\frac{V}{V_s(V-V_s)}$$
$$+\frac{V-V_s}{V}[\eta_{g_1}X_1+\eta_{g_2}X_2]. \tag{4.17}$$

Suffixes 1 and 2 indicate the species 1 and 2 in the mixture, and X_i denotes the mole fraction of the species. The parameters, $Z\kappa$, V_s, E_s, and b are functions of concentration only. The self-diffusion coefficient D may be calculated by means of an equation given by Eyring and Ree (1961):

$$D = \frac{kT}{\xi(V_s/N)^{1/3}\eta} \tag{4.18}$$

where ξ is the number of neighbors of the diffusing molecule in the plane perpendicular to the direction of diffusion. For hexagonal packing, the value of ξ is six. Among many theoretical results, the viscosity and self-diffusion of liquid lithium (Hsu and Eyring, 1972b) are given in Table IX. Recently, starting from the viscosity expression in Eq. (4.14), Faerber *et al.* (1970) obtained the following:

$$\ln\frac{\eta_g}{\eta} = V_s B\left[\frac{1}{(V_g-V_s)T_g} - \frac{1}{(V-V_s)T}\right] + \ln\left[\frac{V-V_s}{V_g-V_s}\left(\frac{T_g}{T}\right)^{1/2}\right] \tag{4.19}$$

where η_g and V_g are the viscosity and the molar volume at the glass transition, respectively, and T_g is the glass transition temperature. B is the constant for the given liquid. It is exciting to see that Eq. (4.19) is able to predict the viscosities at the glass transition region.

Table IX

VISCOSITY AND SELF-DIFFUSION COEFFICIENT OF LIQUID LITHIUM[a]

T (°K)	(cm³/mol)	$\eta_{\text{Calc.}}$ (cp)[b]	$\eta_{\text{Obs.}}$ (cp)	$D_{\text{calc.}} \times 10^4$ (cm²/sec)
500	13.55	0.761	—	0.537
600	13.82	0.508	0.412	0.962
700	14.08	0.418	0.382	1.36
800	14.35	0.361	0.343	1.79
900	14.63	0.320	0.315	2.26
1000	14.91	0.294	0.294	2.73
1100	15.20	0.272	0.278	3.23
1200	15.49	0.257	0.265	3.72
1300	15.78	0.245	0.255	4.20
1400	16.08	0.235	0.247	4.71
1500	16.39	0.225	0.239	5.25
1600	16.69	0.219	0.233	5.74
1623 (Th)[c]	16.76	0.217	0.232	5.85

[a] From Hsu and Eyring (1972b).
[b] cp = Centipoises.
[c] Boiling temperature.

G. THERMAL CONDUCTIVITY

Among many liquid models, the significant structure model also forms the basis for an interpretation of thermal conductivity. Just as in the case of the viscosity, we can write the thermal conductivity of liquid (Lin *et al.*, 1964) as follows:

$$K = (V_s/V)K_s + [(V - V_s)/V]K_g. \tag{4.20}$$

In general, the contribution of gaslike molecules, K_g, is approximated by the equation for the thermal conductivity of an ideal gas (Loeb, 1927)

$$K_g = \frac{(9\gamma - 5)}{4} \eta_g C_{vg} \tag{4.21}$$

where γ is the ratio of the specific heat at constant pressure to that at constant volume, and C_{vg} is the heat capacity at constant volume of a gas. To express K_s, phonon theory Peierls (1929), Kiemens (1951), Callaway, (1959) is used

$$K_s = \frac{C_s^2 R}{V_s(A\omega_e^4 + BT\omega_e^2)} \left[\frac{(\hbar\omega_e/kT)^2 \exp(\hbar\omega_e/kT)}{(\exp(\hbar\omega_e/kT) - 1)^2} \right] \tag{4.22}$$

where ω_e is the Einstein characteristic frequency, $A\omega_e^4$ represents the scattering by lattice imperfections; the term $BT\omega_e^2$ includes the normal and Umklapp processes, and C_s is the velocity of sound in the solid. For the relative high temperature of interest, the term in the square bracket approaches unity.

V. Surface Properties

A. Introduction

So far, we have discussed the treatment of the properties of bulk molecules. However, the situation is quite different for molecules at and near the surface, since they are subject to different forces from the molecules in bulk. It is generally understood that the liquid vapor phase boundaries are not abrupt, but rather that there are transitional layers. Other interesting surface properties such as the adsorption of gases on solid are also characterized by an asymmetry of intermolecular forces. Many models of the liquid have been applied to explain these properties.

B. The Two-Dimensional Liquids

Very few models of two-dimensional liquids have been developed. Devonshire (1937) extended the Lennard-Jones and Devonshire cell theory of liquids to the two-dimensional case and concluded that the critical temperature of the two-dimensional fluid was about half that of the bulk liquid. Recently, Ree *et al.* (1964a) introduced Eyring's idea of "the degeneracy factor" into Ono's (1947) hole theory. This improved hole theory is applied to the two-dimensional case also (Wang *et al.*, 1965). Wang *et al.* (1965) also applied the significant structure theory of liquids to describe a two-dimensional liquid of hard disks. Since Alder and Wainwright (1962) and Alder *et al.* (1963) published exact data for a hard disk system, obtained from molecular dynamics calculations, this system is a good test case for the two-dimensional liquid model. We will now describe the above methods and make comparisons.

1. The Cell Model

Early, Devonshire (1937) extended the cell theory to the case of a two-dimensional (6-12) liquid. Here, we compare the cell theory to the simple fluid of hard spheres (Wang *et al.*, 1965). The following expressions for the partition function, equation of state, and equation of excess entropy are derived.

$$f' = [(2\pi mkT/h^2)a_f]^N(e/3)^N \tag{5.1}$$

$$\pi A/RT = A(\partial \ln a_f/\partial A)_T \tag{5.2}$$

$$S^E/R = \ln(\pi A/RT) + \ln(a_f N/3A). \tag{5.3}$$

The factor $(e/3)^N$ in Eq. (5.1) is artificially included to make the entropy at the limit at $A_0/A = 0$ equal to that of a perfect gas, since at this limit a_f is equal to $3A/N$. This is the well-known communal entropy factor. The "free area" is given by the following equations:

$$a_f = 12(\tfrac{1}{4}a - \tfrac{1}{2}\sigma^2\theta) \qquad \text{if} \quad \sigma > \tfrac{1}{2}a \tag{5.4a}$$

$$a_f = 6(\tfrac{1}{4}3^{1/2}a^2 - \tfrac{1}{3}\pi\sigma^2) \qquad \text{if} \quad \sigma < \tfrac{1}{2}a \tag{5.4b}$$

where σ is the diameter of hard disks and a is the distance between cell centers (see Fig. 11). Here, C and σ are given by

$$C = \tfrac{1}{2}[3^{1/2}a - (4\sigma^2 - a^2)^{1/2}] \tag{5.5a}$$

$$\theta = \cos^{-1}[(a^2 + \sigma^2 - c^2)/2a\sigma]. \tag{5.5b}$$

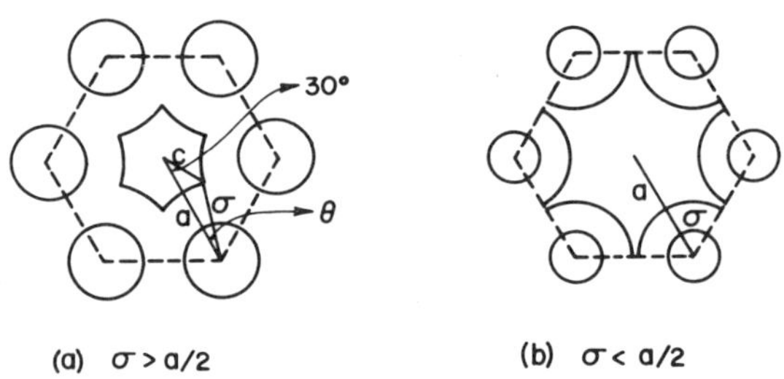

Fig. 11. Free areas in close packing: (a) the case of $\sigma > a/2$: the central hexagon shown by full curves is the free area in this case. (b) the case of $\sigma < a/2$: The hexagon shown by full curves is the free area.

Using the relation $A_0/A = (\sigma/a)^2$, and Eqs. (5.3) to (5.5b), the excess entropies are calculated. Also, obtaining the explicit form for the a_f, the compressibility factor from Eq. (5.2) is obtained. In these calculations, the relation $a_f = 0.05552\ a^2$ is used.

2. The Improved Cell Model

Ono's hole theory is modified by introducing the "degeneracy factor" in the significant liquid structure theory. Then the partition function of the two-dimensional form is given as

$$f' = \left[\frac{2\pi mkT}{h^2}\frac{A}{N}\right]^N \left[\left(1 + Z\frac{A - A_s}{A}\right)\frac{a_f N}{A_s}\right]^{A_s N/A} \times \left[1 - \frac{A_s}{A}\right]^{[1-(A/A_s)]N} \tag{5.6}$$

or putting the values of $z = 6$ and $Na_f/A_s = 0.05552/(\sqrt{3}/2) = 0.06411$, one obtains

$$f' = \left[\frac{2\pi mkT}{h^2}\frac{A}{N}\right]^N \left[0.06411\left(7 - 6\frac{A_s}{A}\right)\right]^{A_sN/A} \left[1 - \frac{A_s}{A}\right]^{[1-(A_s/A)]N}. \quad (5.7)$$

From Eq. (5.7), the compressibility factor and excess entropy are easily obtained, i.e.,

$$\begin{aligned}\frac{\pi A}{RT} = & -1.312\frac{A_0}{A}\ln 0.06411\left(7 - 7.872\frac{A_0}{A}\right) \\ & + \frac{10.328(A_0/A)^2}{7 - 7.872(A_0/A)} \\ & - \frac{A}{1.312A_0}\ln(1 - 1.312A_0/A) \end{aligned} \quad (5.8a)$$

$$\begin{aligned}\frac{S^E}{R} = & \ln\frac{\pi A}{RT} + 1.312\frac{A_0}{A}\ln 0.06411\left(7 - 7.872\frac{A_0}{A}\right) \\ & + \left(1 - \frac{A}{1.312A_0}\right) \\ & \times \ln\left(1 - 1.312\frac{A_0}{A}\right) - 1. \end{aligned} \quad (5.8b)$$

Here $A_s = 1.312A_0$ has been introduced.

3. The Significant Liquid Structure Model

If a perfect-gas partition function for f'_g and a Lennard-Jones and Devonshire partition function for f'(solid) are used, one obtains (Wang *et al.*, 1965) the following partition function for a two-dimensional liquid:

$$\begin{aligned} f' = & \left\{\frac{2\pi mkT}{h^2} a_f \exp\left(-\frac{Z\phi(0)}{2kT}\right)\left[1 + n'_h \exp\left(-\frac{\varepsilon'_h}{RT}\right)\right]\right\}^{A_sN/A} \\ & \times \left\{\frac{2\pi mkT}{h^2}\frac{eA}{N}\right\}^{[(A-A_s)N]/A}. \end{aligned} \quad (5.9)$$

Here every quantity is defined analogously to the quantities in the three-dimensional case. The number of nearest neighbors z is equal to six for close packing; the free area of a molecule a_f is given by

$$a_f = \int_{\text{cell}} \exp\left\{-\frac{z[\phi(r) - \phi(0)]}{2kT}\right\} dr \quad (5.10)$$

where $\phi(0)$ is the potential energy when a molecule of the system is at its two-dimensional cell center.

Using the hard disk conditions,

$$\phi(r) = 0 \qquad r > \sigma$$

$$\phi(r) = \infty \qquad r < \sigma$$

gives $\varepsilon_h' = 0$ since $\varepsilon_h' = a'\phi(0)A_s/(A - A_s)$. Hence, Eq. (5.9) simplifies to

$$f' = \left[\frac{2\pi mkT}{h^2}\frac{eA}{N}\right]^N \left[a_f(1 + n_h')\frac{N}{eA}\right]^{N(A_s/A)}. \tag{5.11}$$

Again, using the value of $A_s = 1.312A_0$ and the free area $a_f = (a - \sigma)^2 = 0.05552a^2$, and making the approximation $n_h' = 6(A - A_s)/A_s$, we can now rewrite Eq. (5.11).

$$f' = \left[\frac{2\pi mkT}{h^2}\frac{eA}{N}\right]^N \left[0.02358\left(6 - 5\frac{A_s}{A}\right)\right]^{A_sN/A}. \tag{5.12}$$

From Eq. (5.12), the compressibility factor $\pi A/RT$ is given as

$$\frac{\pi A}{RT} = 1 - 1.312\frac{A_0}{A}\ln\left[0.02358\left(6 - 6.560\frac{A_0}{A}\right)\right] + \left(1.312\frac{A_0}{A}\right)^2 \frac{5}{6 - 6.56(A_0/A)}. \tag{5.13}$$

And also, the excess entropy S^E, over that of a perfect gas at the same temperature and pressure is readily shown to be

$$\frac{S^E}{R} = \ln\frac{\pi A}{RT} + 1.312\frac{A_0}{A}\ln\left[0.02358\left(6 - 6.56\frac{A_0}{A}\right)\right]. \tag{5.14}$$

In deriving Eq. (5.14), the partition function for an ideal two-dimensional gas has been taken as

$$f_g'(\text{ideal}) = \left[\left(\frac{2\pi mkT}{h^2}\right)\left(\frac{eA_{id}}{N}\right)\right]^N \tag{5.15}$$

and the relation $\pi A_{id}/RT = 1$ has been substituted.

On the other hand, using molecular dynamics, the excess entropy is calculated by means of the following equation (Wang *et al.*, 1965):

$$\frac{S^E}{R} = \ln\frac{\pi A}{RT} + \int_\infty^{A/A_0}\left[\frac{\pi A}{RT} - 1\right]\frac{A_0}{A}\,d\left(\frac{A}{A_0}\right). \tag{5.16}$$

In Table X, the excess entropies and the compressibility factors obtained by the molecular dynamics method (MD), the significant structure theory (SST), the cell theory (CT), and the improved hole theory (IHT) are compared. One sees that SST and IHT show a definite improvement over the original cell theory. The common features of SST and IHT, which include some kind of degeneracy factor, provide a suggestion for improving the liquid model.

Table X

COMPARISON OF THE THERMODYNAMIC DATA OF A HARD-DISK LIQUID CALCULATED FROM VARIOUS THEORIES WITH MACHINE CALCULATIONS[a]

	S^E/R					$\pi A/RT$				
A_0/A	MD	SST	CT	SPT	IHT	MD	SST	CT	SPT	IHT
0.4	0.13	0.33	0.59	0.12	0.11	2.49	2.75	2.85	2.46	2.58
0.5	0.24	0.53	0.86	0.23	0.24	3.41	3.61	3.51	3.35	3.54
0.6	0.42	0.81	1.15	0.41	0.47	5.03	4.90	4.53	4.81	5.11
0.7	0.80	1.18	1.47	0.73	0.82	7.73	7.15	6.57	7.50	8.28
0.8	1.51	1.73	1.95	1.35		11.99	12.58	9.53	13.28	

[a] From Eyring and Jhon (1969, p. 128).

[b] MD, Molecular dynamics; SST, Significant-structure theory; CT, Cell theory; SPT, Scaled particle theory; IHT, Improved hole theory.

C. MODELS OF PHYSICAL ADSORPTION

Near the boundary between solid and gas phases, it is usually true that the surface density is several orders of magnitude greater than that in the gas phase. Hence, it is quite reasonable to treat the intermediate region near the surface as a separate adsorbed phase. If the forces holding molecules near the surface are of a physical nature such as a van der Waals force, we speak of "physical adsorption," while if the forces are of a chemical nature, i.e., involve chemical bonding, we speak of "chemical adsorption." Among models used to understand the problem of physical adsorption of gases on solids, the significant structure model and the improved cell model have been successfully developed. One can start with partition functions describing the adsorbed and bulk phases and equate the chemical potential of the two phases. Using this relation, many properties can be evaluated.

1. *Application to the Monolayer Region*

First, we consider a simplified system whose surface is a structureless uniform plane with physical adsorption occurring in the submonolayer region. McAlpin (1963) and McAlpin and Pierotti (1964) applied the significant liquid structure model to the physical adsorption of Ar and Kr on a graphite surface. According to their model, the adsorbed phase is a two-dimensional fluid adsorbed on a solid that possesses an energetically homogeneous plane surface. The molecules of the fluid interact with the solid and vibrate normal to the surface of the solid. The interaction is assumed to fall off rapidly enough with distance (Pierotti, 1962) to ensure that at low temperatures only the first layer nearest the surface will be populated (only the monolayer adsorption is considered). Preserving the form of Eq. (2.1), one can write the partition function for a two-dimensional adsorbed fluid as

$$f_{\mathrm{ads}} = \{f_{2\mathrm{s}} f_{2\mathrm{deg}}\}^{N(A_{\mathrm{s}}/A)}[\{f_{2\mathrm{g}}\}^{[N(A - A_{\mathrm{s}})/A]}] \tag{5.17}$$

where f_{ads}, $f_{2\mathrm{s}}$, and $f_{2\mathrm{g}}$ are the partition functions for the adsorbed phase, the two-dimensional solidlike structure including the adsorbate–absorbent interaction terms, the positional degeneracy, the two-dimensional gaslike structure including the adsorbate–adsorbent interaction terms, respectively. As in the molar area of the solid adsorbate, A is the molar area of the adsorbed phase at a given surface coverage θ, and N is the total number of molecules in the adsorbed phase. Next, we consider the relation

$$A = A_{\mathrm{m}}^{0}/\theta \tag{5.18}$$

where A_{m}^{0} is equal to $LA_{\mathrm{m}}/N_{\mathrm{m}}$, A_{m} is the area of the surface, N_{m} is the number of molecules on the surface at $\theta = 1$, respectively, and L is Avogadro's number. Substitution of these quantities into Eq. (5.17) and recalling the relation $N = \theta N_{\mathrm{m}}$ yields

$$f_{\mathrm{ads}} = (f_{2\mathrm{s}} f_{2\mathrm{deg}})^{\theta 2 N_{\mathrm{m}} A_{\mathrm{s}}/A_{\mathrm{m}}^{0}} (f_{2\mathrm{g}})^{\theta N_{\mathrm{m}}(1-\theta)}. \tag{5.19}$$

The chemical potential for the adsorbed phase and the gas phase are

$$\mu_{\mathrm{ads}} = \left(\frac{\partial F_{\mathrm{ads}}}{\partial N}\right)_{A_{\mathrm{m}}, T} \tag{5.20}$$

where $F_{\mathrm{ads}} = -kT \ln f_{\mathrm{ads}}$ and

$$\mu_{\mathrm{gas}} = \mu_{\mathrm{gas}}^{0} + kT \ln p. \tag{5.21}$$

Equating Eqs. (5.20) and (5.21), an isotherm is obtained. The isosteric heat of adsorption q_{st} is given by

$$q_{\mathrm{st}} = RT^2(\partial \ln p/\partial T)_\theta . \tag{5.22}$$

The critical coverage θ_{c} and the critical temperature T_{c} can be determined using the following conditions:

$$(\partial p/\partial \theta)_{T, N_{\mathrm{m}}} = 0, \qquad (\partial^2 p/\partial \theta^2)_{T, N_{\mathrm{m}}} = 0. \tag{5.23}$$

The critical pressure can be obtained from θ_{c}, T_{c}, and the isotherm. The second virial coefficient B_2 has the value

$$B_2 = \left[\frac{\partial[\pi A/RT]}{\partial(1/A)}\right]_{1/A\to 0} \tag{5.24}$$

where $\pi = -(\partial F_{\text{ads}}/\partial A)_T$.

In the actual application of the model to the physical adsorption of Ar and Kr on a graphite surface (McAlpin and Pierotti, 1964), the following expressions for f_{2s}, $f_{2\text{deg}}$, and f_{2g} in Eq. (5.19) are used. f_{2s} can be expressed as the product of the partition function of a molecule in a two-dimensional Einstein crystal with θ_E as the Einstein temperature, and a partition function for the vibration normal to the surface. Then,

$$f_{2s} = \left(\frac{\exp(-\theta_E/2T)}{1-\exp(-\theta_E/T)}\right)^2 \exp\frac{U_0 + W}{RT}\left(\frac{\exp(-h\nu_0/2kT)}{1-\exp(-h\nu_0/kT)}\right). \tag{5.25}$$

Here U_0, W, and ν_0 are the molecular adsorbate–adsorbent interaction energy, the molar two-dimensional lattice energy of the adsorbate, and the frequency of the vibration, respectively. The second term in Eq. (5.25) is a correction to the standard energy. Analogously, f_{2g} is given by

$$f_{2g} = \left(\frac{2\pi mkT}{h^2}\right)\frac{eA}{L}\exp\frac{U_0}{RT}\left(\frac{\exp(-h\nu_0/2kT)}{1-\exp(-h\nu_0/kT)}\right) \tag{5.26}$$

where the first two terms represent the partition functions of the two-dimensional ideal gas times e to account for the indistinguishability of the molecules and the last two terms represent the adsorbate–adsorbent interaction as mentioned above.

As pointed out in Section II, $f_{2\text{deg}}$ can be expressed as

$$1 + n_h \exp(-\varepsilon_h/RT),$$

and ε_h is an inverse function of the fraction of nearest neighbor vacancies. In normal liquids, this fraction is approximately 0.1, whereas in the adsorbed phase even at θ as high as 0.7, the fraction of nearest-neighbor vacancies is three times as high as in a normal liquid. Hence, it is assumed that ε_h is negligible in the calculation; and n_h is given as $z(1-\theta)$, where z is the number of nearest neighbors. (For a hexagonal lattice, as assumed here, $z = 6$.) The two-dimensional lattice energy W was obtained by summing the Sinaňoglu and Pitzer (1960) modification of the Lennard-Jones (6-12) potential over the nearest 18 neighbors in the hexagonal lattice and integrating over the rest of the lattice. The Einstein characteristic temperature θ_E was obtained by multiplying the ratio of the perturbed and unperturbed frequencies determined from the curvature of the potential energy curves by the Einstein temperature of the bulk adsorbate. Quantities such as U_0, W, ν_0, and A_m^0 are obtainable from the references given.

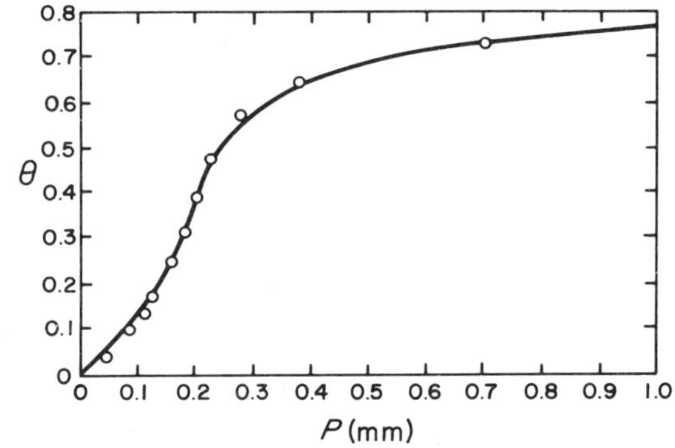

Fig. 12. Argon–graphite isotherm at 77.6°K for Argon p-33. Solid line is theoretical. Points are experimental.

Some of the calculated adsorption isotherms for argon and krypton adsorbed on graphite are compared with experiment in Figs. 12 and 13. The agreement for argon adsorbed on graphite is excellent at low and intermediate coverage. The agreement for Kr on graphite is again very good, but not as good as with argon. This may be due to the few percent error in W, which was calculated as described above. For krypton, there was no experimental verification of the third-order dispersion effect as there was in the case of argon. The theoretical isosteric heat of adsorption of argon on graphite at 84° and 150°K was calculated using Eq. (5.22) and the results are shown plotted as a function of θ in Fig. 14. These values are compared with the experimental heats determined by Ross and Oliver (1961) for argon on graphite at 84°K. The agreement is quite good.

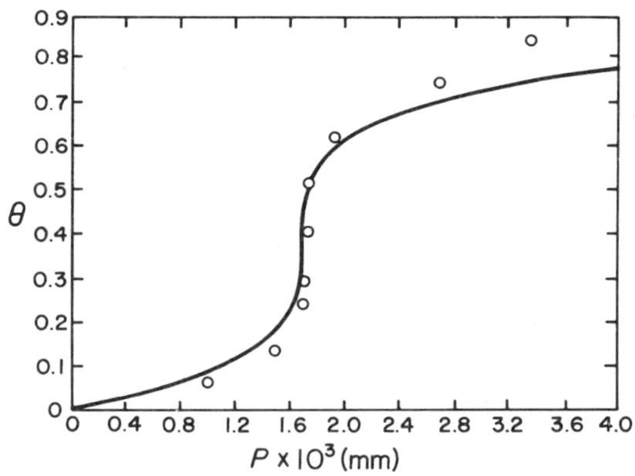

Fig. 13. Krypton–graphite isotherm at 77.8°K for Kr on p-33. Solid line is theoretical. Points are experimental.

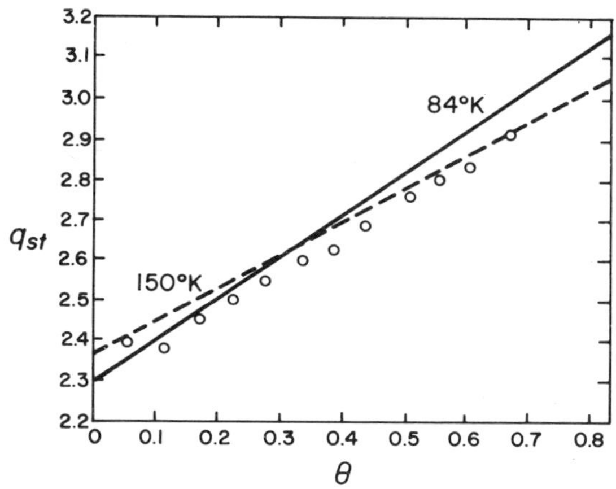

Fig. 14. Isosteric heats of adsorption (kilocalories per mole) for Ar on p-33 versus coverage for argon on graphite. The lines are theoretical. Points are experimental.

There has been much discussion about the slopes of heat versus coverage curves (Young and Crowell, 1962). At 84°K, the slope of the experimental curve is 1000 cal/mole while the theoretical one is 1050 cal/mole. Sams *et al.* (1962) reported a slope for this system of 810 cal/mole, while that of the theoretical curve at 150°K is 833 cal/mole. θ_c for the present model is found to be 0.3824, whereas that of a two-dimensional van der Waals fluid is 0.333 and that of a localized Bragg–Williams monolayer is 0.5. The critical temperature for argon on graphite is predicted to be 65°K, whereas the experiment is approximately 68°K (Prenzlow and Halsey, 1957). Thus, the significant liquid structure model is quite useful in explaining the many properties of physical adsorption.

However, it is not possible to obtain a two-dimensional virial coefficient from the present model. In order to get second virial coefficients, the appropriate modification of Eyring's expression, $1 + n((V - V_s)/V_s)\exp(-\varepsilon_h/RT)$, for the three-dimensional liquid is required. In the case where $\varepsilon_h = 0$, this expression for a two-dimensional liquid becomes $1 + n(1 - \theta)/\theta$. If the model discussed above is to be practically useful, it must be applicable to more complex molecules than monatomic gases. These molecules involve rotational and vibrational partition functions, in addition to the simple partition function for the adsorbed atoms. From the available data (Hoory and Prausnitz, 1967), restricted rotation is an important aspect of the theory of physical adsorption for complex molecules. The hindered or restricted rotation model is tested for physical adsorption on graphite of the gases, N_2, $CHCl_3$, CCl_4 (Wang *et al.*, 1971) and hydrocarbons (Kim *et al.*, 1976) with good success.

The significant liquid structure model is also extended to physical adsorption of quantum gases (Kim *et al.*, 1972) such as H_2, D_2, and He, which exhibit quantum effects due to their small masses and the small forces involved. In formulating the partition function, the Debye model for a two-dimensional quantum gas with weak degeneracy and spin degeneracy is introduced for the gaslike partition function, which follows either Bose–Einstein statistics or Fermi–Dirac statistics. The derivations of these expressions are analogous to those for three-dimensional quantum liquid which have already been discussed in Section IV (Henderson *et al.*, 1962). The calculated results for the adsorption isotherms and for the isoteric heat give good agreement with experiment.

2. Application to the Multilayer Region

Most of the recent theoretical work in physical adsorption has been carried out for the submonolayer region with low or intermediate values of surface coverage. There have been few investigations in the multilayer region, but one such formulation has been developed by Pace (1957) using Ono's free volume model.

According to this model, the observed differential or isosteric heat of adsorption is explained for argon on graphite at concentrations between 0.5 and 2.3 monolayers for temperatures in the neighborhood of the normal boiling point of argon. The assumption was made that the adsorption proceeds in a layerwise fashion with the adsorbed molecules in each layer exhibiting a behavior that can best be described as liquidlike. The energy of a molecule on the surface is considered to be the sum of two contributions: (1) a site energy for the isolated molecule and (2) an interaction energy with neighbors in the same layer. The site energy for a molecule is calculated assuming atoms and molecules in lower layers are fixed with respect to both a configuration in a hexagonally packed planar lattice and in the distance between layers. The interaction energy is calculated assuming that the behavior of the adsorbed gas in any layer can be represented by a two-dimensional lattice or free volume model. Here, the Ono (1947) approximation has been used to express the variation of the free volume with the fraction of vacant sites in the lattice. The assembly partition function f for i adsorbed layers can now be written as

$$f = \prod_i \{\exp[(-N_i/kT)(X_i + z\mu_c/2)]\}\{(2\pi mkTJ_0/h^2)^{N_c}\} \times \left\{\sum_{N_{i0}} g(N_i, N_{i1}, N_{i0})[\exp(-\psi_i/kT)]^{N_a/2}\right\} \tag{5.27}$$

in which

$$\psi_i = -[\mu_i + (2kT/C)\ln(J_1/J_0)]. \tag{5.28}$$

Here, X_i is the site energy for the isolated molecule in the ith layer referred to an energy zero for the molecule at infinite separation from the surface at absolute zero, z is the coordination number of the lattice, and μ_i is the interaction energy for the equilibrium distance between nearest neighbors. The parameters J_0 and J_1 in the Ono approximation are the values of the generalized free volume J_w setting $w = 0$ and 1, respectively; w is the fraction of sites vacant around any lattice site. $f(N_i, N_{i-1}, N_{i0})$ denotes the number of arrangements of N_i molecules in the ith layer on the N_{i-1} sites in the $(i - 1)$th layer to give N_{i0} "molecule–hole" pairs on adjacent sites in the ith layer. To obtain the value of N_{i0}, the distribution of adsorbed molecules in any layer is considered to be completely random. The calculated isosteric heats of adsorption for the system Ar-graphitized carbon black give good agreement with the experiment. Although Pace did not derive any specific theoretical isotherms, Pierotti and Thomas (1971) have shown that the isotherms predicted by Pace's theory are not accurate representations of the actural experimental systems.

Very recently, a new model of multilayer physical adsorption (Newsome, 1974) has been developed using a modified form of the significant structure theory of liquids and the general approach of Pace. The total partition function f for all i adsorbed layer is taken as

$$f = \prod_i f_i = \prod_i q_{is}^{N_i(N_i/N_{i-1})} q_{ig}^{N_i[(N_{i-1} - N_i)/N_{i-1}]} \tag{5.29}$$

where f_i is the partition function for the ith adsorbed layer, q_{is} is the partition function for the solidlike structure in the ith layer and q_{ig} is the corresponding partition function for the gaslike structure. N_i is the number of molecules adsorbed in the ith layer, while N_{i-1} is the number of molecules adsorbed in the $(i - 1)$th layer. N_0 corresponds to the number of sites available for adsorption on the surface. For q_{is}, the partition function for the two-dimensional cell model as given by Devonshire (1937) is used in this development, with a term for the configurational degeneracy and the strain energy assumed to be zero. Thus,

$$q_{is} = (\Lambda^{-2} a_{fi}) \left[\exp\left(\frac{U_i + W_i}{RT} \right) \right] (1 + n_h) q_z \tag{5.30}$$

where Λ is $h/(2\pi mkT)^{1/2}$, a_{fi} is the free area in the ith layer, U_i is the molar site energy for an isolated molecule in the ith layer, and W_i is the molar two-dimensional lattice energy of the adsorbate in the ith layer. For this model, n_h is given as

$$n_h = z(A_i - A_i^0)/A_i^0 \tag{5.31}$$

where z is the coordination number, A_i is the area per molecule in the ith layer, and A_i^0 is the area per molecule when the ith layer is filled to capacity. The harmonic oscillator partition function for the vibration normal to the adsorbing surface, q_z, is assumed to be the same for all layers and is given by

$$q_z = \frac{\exp(-h\nu/2kT)}{[1-\exp(-h\nu/kT)]} \tag{5.32}$$

where ν is the vibrational frequency. For q_{ig}, the two-dimensional ideal gas model is used.

$$q_{ig} = q_z[\exp(\mu_i/kT)](\Lambda^{-2}A_i e). \tag{5.33}$$

Theoretical isotherms are obtained by finding the chemical potential for the adsorbed molecules in each layer and equating these to each other and to the chemical potential of the gas phase. The total statistical coverage θ is defined as

$$\theta = \frac{1}{N_0}\sum_{i-1} N_i = \frac{1}{N_0}\sum_{i=1} \theta_i N_{i-1}. \tag{5.34}$$

Here θ_i is the coverage in the ith layer. The model seems to provide an excellent fit of the observed data of Ar–BN system in Fig. 15. At the highest

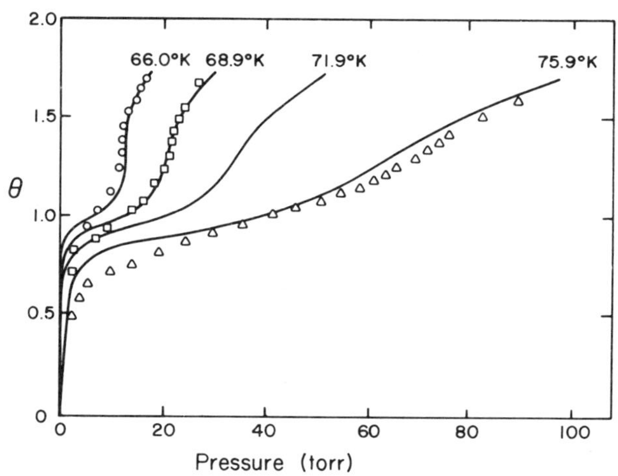

Fig. 15. Theoretical isotherms and experimental data for Ar on boron nitride.

temperature, there is no step in the data or in the theory, but as the temperature is lowered, both the theory and data show the development of a well-defined step in the second layer. At the lowest temperature of interest, both theory and experiment show condensation.

D. Surface Tension of Pure Liquids

With the use of various liquid models, several theories have been proposed for the surface tension of liquids. Among them, for a pure liquid, the model of the significant liquid structure is most successful and is most widely applied (Eyring and Jhon, 1969). The lattice theory or cell theory for mixtures has been studied for a considerable time and has been developed in various ways (Guggenheim, 1952; Defay *et al.*, 1966). Recently, the significant liquid structure model has been applied to mixtures.

There are two methods for obtaining the surface tension. One is an iteration method (Chang *et al.*, 1962), which calculates the sum of the contributions of successive layers, and the other is a monolayer approximation originally developed using a cell model (Prigogine and Saraga, 1952) and later used in a significant liquid structure model (Ree *et al.*, 1964c; Lu *et al.*, 1967). The former procedure is more faithful to physical reality, but the latter requires fewer data for calculation. Using one of these two methods, the theory has been successfully applied to simple liquids (Chang *et al.*, 1962; Ree *et al.*, 1964c), polar liquid (Lee *et al.*, 1964; Lu *et al.*, 1967), molten metals and salts (Jhon *et al.*, 1968; Hsu and Eyring, 1972a; Pack *et al.*, 1970), liquid crystals (Do *et al.*, 1976), and polyethylene (Stewart and von Frankenberg, 1968; Ma *et al.*, 1974). Using the iteration method, one finds there is an appreciable density change over the first two or three layers.

The surface tension of a liquid γ is the excess Gibbs free energy per unit area over that of the bulk liquid. Accordingly, the surface tension is given by

$$\gamma = \sum_i (G_{si} - G_l)\frac{d}{V_i} = \sum_i (G_{si} - G_l)\frac{0.9165(V_s/N)^{1/3}}{V_i} \tag{5.35}$$

where G_{si} is the molar Gibbs free energy of the ith surface layer: G_l the molar Gibbs free energy of the bulk liquid; V_i the molar volume of the ith layer and the thickness d of a molecular layer which can be deduced from the cubic closest packing structure as $0.9165(V_s/N)^{1/3}/V_i$. For example, in one procedure the molar free energy of the ith surface layer for monatomic liquids is obtained using the bulk partition function [Eq. (2.7)], by substituting the appropriate values of E_{si}, the sublimation energy of a molecule, in the ith layer. The correction of E_s, the energy of a molecule in the ith surface layer E_{si} is calculated using the following equation,

$$E_{si} = E_s\left(\frac{6}{12}\frac{\rho_i}{\rho_i} + \frac{3}{12}\frac{\rho_{i+1}}{\rho_i} + \frac{3}{12}\frac{\rho_{i-1}}{\rho_i}\right) \tag{5.36}$$

where ρ_i, ρ_{i+1}, and ρ_{i-1} are the densities of the ith layer and the layers immediately below and above. For a simple close-packed liquid, six-twelfths of the neighbors lie in the same layer, three-twelfths in the layer below, and three-twelfths in the layer above. The corresponding molar volume V_i is determined by plotting A_i versus $X_i = V_i/V_s$, and by drawing a tangent whose slope is the vapor pressure; then V_i is the volume at the tangential point (see Section IV). In the actual calculation of $V_i(\rho_i)$ and E_{si}, the iteration is repeated. In Table XI, the calculated surface tensions of argon (Chang *et al.*, 1962) obtained by the above scheme are compared with the observed values. The results are quite satisfactory. Unlike simple liquids, polar liquids such as water, alcohol, and liquid ammonia (Lee *et al.*, 1964; Lu *et al.*, 1967)

Table XI

MOLECULAR PARAMETERS AND CONSTANTS FOR Ar AND Kr ON GRAPHITE[a]

	Ar	Kr
U_0 (cal/mole)	2220	2860
ν_0 ($\sec^{-1}$)	1.20×10^{12}	1.00×10^{12}
A_m^0/L (Å^2/molecule)	14.8 (77.6°K)	19.5 (78°K)
ε/k (°K)	119.8	171
σ (cm)	3.405	3.60
W (cal/mole)	710	890
θ_E (°K)	45.0	33.8
V_m (cc (S.T.P)/g)	3.66	2.86

[a] From Eyring and Jhon (1969, p. 132).

tend to orient their dipoles with respect to the surface. When account is taken of this, satisfactory agreement with experiment is obtained.

The iterative approach fails to provide a simple equation for calculating the surface tension, since to derive an equation for the surface tension in closed form Prigogine and Saraga (1952) assumed that the dividing surface between a liquid and its vapor phase is a monomolecular layer in which a molecule has a free volume larger than an interior molecule and a potential energy less than the latter. This is a fair approximation for simple liquids, but is an oversimplification for a hydrogen-bonded associated liquid such as water (Jhon *et al.*, 1967).

According to the closed expression, the surface tension γ is given by the Helmholtz free energy A as

$$\gamma = \left(\frac{\partial A}{\partial \Omega}\right)_{N,V,T} = w^{-1}\left(\frac{\partial A}{\partial N_c}\right)_{N,V,T} \tag{5.37}$$

where Ω is the surface area and N_0, the total number of sites available on the surface, is related to N by the relation $N' = (V_s/V)N_0$, since a random distribution of holes is assumed. Here, $N = N' + N''$; N' and N'' are the number of molecules in the surface and in the bulk liquid, respectively.

One of the most successful equations of surface tension was derived in terms of significant liquid structure theory by Ree *et al.* (1964c). According to them,

$$\gamma = \frac{2}{\sqrt{3}}\left(\frac{N}{\sqrt{2}\,V_s}\right)^{2/3}\left(\frac{V_s}{V}\right)^2\left[\frac{z}{2}\psi'(0) - \frac{z}{2}\psi(0) - kT\ln\frac{v_f'}{v_f}\right]. \tag{5.38}$$

Here, $\psi(0)$ is the potential energy possessed by a molecule as it vibrates about its equilibrium position and z is the coordination number of the

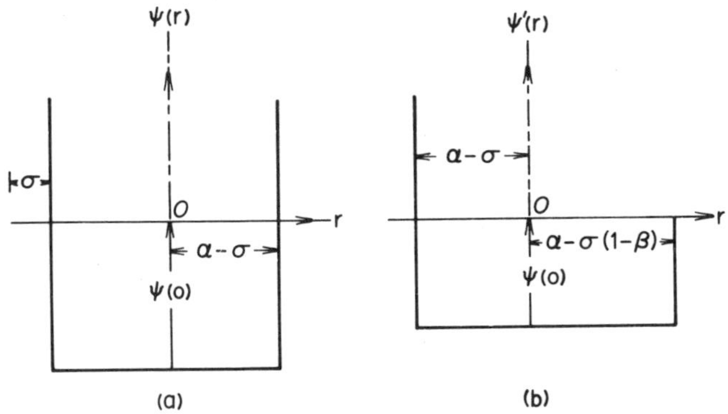

Fig. 16. Square well potential: (a) the potential for an internal molecule; (b) the potential for a surface molecule.

molecule. The potential energy for a surface molecule $\psi'(0)$ is given as $\psi'(0) = 9/12\psi(0)$. To obtain the value of v_f'/v_f and $\psi'(0) - \psi(0)$, they used a square well potential and a Lennard-Jones potential as Prigogine and Saraga (1952) did (see Fig. 16). The width of the square well for the motion toward the bulk liquid is $(a - \sigma)$, and toward the vapor phase it is $a - \sigma(1 - \beta)$. Hence, v_f and v_f' are written

$$v_f = \tfrac{4}{3}\pi(a - \sigma)^3 \tag{5.39}$$

$$v_f' = \tfrac{1}{2}(\tfrac{4}{3}\pi(a - \sigma)^3) + \tfrac{1}{2}(\tfrac{4}{3}\pi(a - \sigma)^2(a - \sigma)(1 - \beta)). \tag{5.40}$$

Here the value of $\beta = 0.25$ is found by Prigogine and Saraga (1952). Defining the reduced quantities, $kT/\varepsilon = T^*$, $V/N\sigma^3 = v^*$, and $(N\sigma^3/V_s) = \alpha$, the reduced surface tension γ^* is compared with those from other theories. One can see that the significant liquid structure approach is most successful. Here, the nonsuperposability of the values for He, H_2, and Ne is due to a quantum effect.

E. Surface Tension of Liquid Mixtures

Among many models of surface tension of liquid mixture, two approaches are most successful. One is the quasi-lattice model founded by Guggenheim (1952) and Defay *et al.* (1966) (see Section IV). The other is the significant liquid structure approach (Kim *et al.*, 1968, 1969). According to the former approach, Guggenheim derived the equation that relates the surface tension of a binary mixture to the activities of its components in the bulk liquid and in the surface. However, this model is only applicable to the mixtures whose components are similar in molecular size and shape. Englert-Chowles and Prigogine (1958a,b) developed the theory by treating the equilibrium surface tension of a mixture as a perturbation on the "dynamic surface tension" using a smoothed potential function. This is only applicable to nearly ideal systems.

Taking into account the size effect of the molecule, Hildebrand and Scott (1950) extended the Guggenheim model. Later, Eckert and Prausnitz (1964) developed the surface tension of nonpolar binary liquids using the cell model and the grand partition function of the surface which is basically an improved Guggenheim equation. All the above efforts are limited in their use, however, the surface tension of regular solutions has been given a firm theoretical foundation by Guggenheim and Prigogine. They have shown that, when each molecule in the liquid is treated as if it were located on a lattice (quasi-lattice model) and there is a random distribution of the molecules in both the bulk and the surface of the liquid (zeroth approximation), in spite of the nonzero heat of mixing, the surface tension of the mixture is given by

$$\begin{aligned}\gamma = \gamma_1 + \frac{kT}{a}\ln\frac{x'_1}{x_1} + \frac{wl}{a}[(x'_2)^2 - (x_2)^2] \\ - \frac{wm}{a}(x_2)^2 = \gamma_2 + \frac{kT}{a}\ln\frac{x'_2}{x_2} \\ + \frac{wl}{a}[(x'_1)^2 - (x_1)^2] - \frac{wm}{a}(x_1)^2 \qquad (5.41)\end{aligned}$$

where γ_i is the surface tension of the pure ith component at the given temperature; x_i is the mole fraction of the ith component; the primes indicate the surface layer; w is the interaction parameter in the theory of regular solutions; l and m refer to the fraction of nearest neighbors that occupy the lattice plane and an adjacent lattice plane, respectively, and satisfy the relation $l + 2m = 1$. For example, for a close-packed lattice $z = 12$ and $m = \frac{1}{4}$. The parameter a, which is the mean surface area of the molecules, is given by

$$a = \tfrac{1}{2}[(v_1/N)^{2/3} + (v_2/N)^{2/3}] \qquad (5.41)$$

where N is avogadro's number and v_i is the molar volume of the ith component.

The quasilattice model has been applied not only to simple binary mixtures but also to fused salt mixtures successfully (Lumsden, 1966; Nissen and Carlsten, 1974). We developed various ways for obtaining the surface tension of pure liquid from the significant structure theory of liquid. The extension of the approach to binary liquid mixtures, such as cryogenic mixtures, is made by using either the iteration method or the monolayer approximation

(Kim *et al.*, 1968) with satisfactory results. However, the basic understanding of the prediction of the activity coefficient of each component is not well developed and application of the method as a result has some limitations.

VI. Properties and Structure of Water and Aqueous Solutions

A. Introduction

In this section we will be chiefly concerned with the development of the liquid model approach in connection with the structure and properties of liquid water and aqueous solutions. In general, the theoretical study for liquids has been extremely difficult due to the complex nature of the intermolecular potential. In particular, for water, with its many degrees of freedom, extensive numerical calculations are nearly impossible. So far, most theoretical studies have dealt with simplified models. All of these treatments suffer from a variety of defects such as severe structural underdefinition and excessive reliance on adjustable parameters. Also, some of the models disagree with the available experimental facts.

Water, the most common liquid on earth, exhibits a variety of unusual properties frequently overlooked or ignored. In the past, many books have been written on the structure and properties of water and on aqueous solutions (Kavanau, 1964; Samoilov, 1965; Eisenberg and Kauzmann, 1969; Horne, 1972; Franks, 1972). The available models for the structure of water can be classified into two categories: (1) uniformist models, which assume that all water molecules are subject to the same intermolecular force interactions and (2) mixture models, which assume the presence of at least two or more structures.

In the past, numerous models have been proposed. However, there are no methods available to study the instantaneous structure of water (Vinogradov and Linnell, 1971). To do this, a technique having a time resolution of 10^{-13} to 10^{-14} sec (the vibration time) or less is necessary. Only a few methods, i.e., infrared and Raman spectroscopy, neutrons in elastic scattering, and dielectric relaxation are available for the vibrationally averaged structures. These require resolution times greater than 10^{-13} sec but less than 10^{-10} sec (the diffusion time). Many commonly available techniques involving observation times greater than 10^{-10} sec provide information on the diffusion-averaged structures. X-Ray diffraction, thermodynamic data, and NMR fall in this category. From Raman absorption studies, Wall and Hornig (1965) concluded that the uniform model gives a satisfactory fit to the data. However, Walrafen's (1968) later experiments support a mixture model. The infrared study of Senior and Verall (1972) agree with Walrafen's data. On the basis of X-ray scattering experiments, Narten *et al.* (1967) and

Narten and Levy (1969) expresses a negative view of mixture models. All mixture models assume the existence of microphase or "clusters" in the liquid. Considering all the above facts, the cluster or microphase lifetime must be less than 10^{-11} sec (Frank and Wen, 1957). The neutron diffraction data of Safford *et al.* (1969) support this view.

The models that successfully account for the varied properties of liquid water involve short-lived clusters of more than one type. We will discuss the several advanced models of this type—the model of Neméthy and Scheraga (1962, 1969), modified by Hagler *et al.* (1973) (i.e., basically a modified version of the significant liquid structure approach); that of Jhon *et al.* (1966, 1967) and Hobbs *et al.* (1966) (i.e., a significant liquid structure approach); and that of Weres and Rice (1971) (i.e., a modified version of cell theory). This discussion will be followed by consideration of the properties and structures of aqueous solutions containing hydrocarbons (Neméthy and Scheraga, 1962, 1969; Herman, 1971; Sung *et al.*, 1972) and ions (Jhon and Eyring, 1968a).

B. Models of Neméthy and Scheraga

1. Original Treatment

Neméthy and Scheraga (1962, 1969) developed models of water akin to the theory suggested by Frank and Wen (1957). Hydrogen-bonded clusters with lifetimes in the neighborhood of 10^{-10} to 10^{-11} sec were postulated. The clusters are considered to be imbedded in and in equilibrium with monomeric "unbonded" water. Neméthy and Scheraga assumed that liquid water consists of a mixture of monomers and hydrogen bonded clusters of uniform size. The five species of water molecules postulated were zero (unbonded), one, two, three, and four hydrogen-bonded molecules (see Fig. 17).

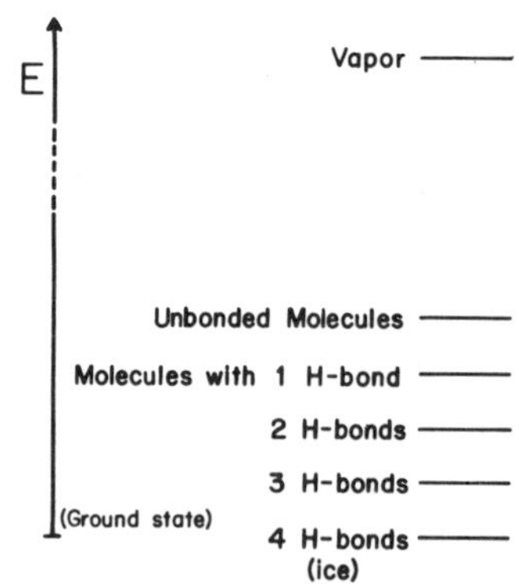

Fig. 17. Schematic representation of energy levels for H_2O molecules in liquid water.

In order to obtain the thermodynamic properties of liquid water as functions of temperature, a partition function similar to that used in "significant liquid structures" was developed. The five species are tested as occupants of five energy levels. They then obtain the following partition function:

$$f = \sum_{\{n_0,\, y_1,\, x_u\}} g(n_0, y_1, x_u)\{f_u \exp(-E_u/RT)\}^{N_0 x_u} \times \left[\prod_{i=1}^{4} \{f_i \exp(-E_i/RT)\}^{N_0 x_i}\right] \tag{6.1}$$

where one has

$$g(n_0, y_1, x_u) = \frac{N_0!}{N_4!\,N_3!\,N_2!\,N_1!\,N_u!}$$

and the N_i are expressed in terms of the cluster size (n_0), the fraction (y_1) of one bonded molecule in a cluster and of x_u. N_0, x_u, and x_i are Avogadro's number, and the mole fractions of the various species. N_u and N_i, E_u and E_i, and f_u and f_i are, respectively, the number of molecules, the energies relative to the four H-bonded species, and the weighting factors of the various species. $N_i = N_0 x_i$ ($i = 4, 3, 2, 1$) is the number of molecules of each species with u referring throughout to the unbonded species. The weighting factors f_i of the various species are given as

$$f_u = \left\{\left(\frac{2\pi mkT}{h^2}\right)^{3/2} v_f(2\pi/\sigma)\left(\frac{2\pi IkT}{h^2}\right)^{1/2}\right\} f_v$$

with

$$f_i = f_v = \prod_{j=1}^{s} [1 - \exp(-h\nu_{ij}/kT)]^{-1} \tag{6.2}$$

with $s = 6$ for $i = 1, 2, 3, 4$ and $s = 2$ for $i = \text{u}$. σ is the symmetry number (for water $\sigma = 2$), n is the mass of the molecule, I is the moment of inertia around the dipole axis, and v_f is the free volume for translation of an unbonded molecule. The other symbols have their usual statistical significance. The energy levels are equally spaced and the energies assigned to the five levels are $E_4 = 0$, $E_3 = 0.5\ E_H$, $E_2 = 1.0\ E_H$, $E_1 = 1.5\ E_H$ and $E_u = 2.0\ E_H$. Here, the hydrogen bond energy (E_H) and the free volume (v_f) are also adjustable parameters in the model. The values taken by the three parameters n_0, y_1, x_u must satisfy the condition of minimum free energy for the system. These values can be found by setting $A = -kT \ln f$, using Stirling's approximation, and finally solving the following simultaneous equations:

$$\partial A/\partial n_0 = 0 \qquad \partial A/\partial y_1 = 0 \qquad \partial A/\partial x_u = 0. \tag{6.3}$$

Using the above fitting technique, many thermodynamic properties for H_2O as well as D_2O were successfully calculated. In these calculations, the parameters in Table XII were used; as indicated in columns 3, 4, and 6 of the

Table XII

MOLAR VOLUMES AND VAPOR PRESSURES OF WATER[a]

T(°K)	V_{calc} (cm^3)	V_{obs} (cm^3)	% error	P_{calc} (atm)	P_{obs} (atm)	% error
273.15 (m.p.)	17.923	18.019	−0.53	0.006051	0.006030	−0.35
277.15	17.888	18.016	−0.71	0.008036	0.008007	−0.36
283.15	17.891	18.021	−0.72	0.01210	0.01218	−0.66
293.15	17.961	18.048	−0.48	0.02298	0.02307	−0.39
313.15	18.171	18.157	0.08	0.07234	0.07279	−0.62
353.15	18.673	18.538	0.73	0.4623	0.4672	−1.05
373.15 (b.p.)	18.959	18.799	0.85	0.9852	1.0000	−1.48
423.15	19.800	19.641	0.81	4.549	4.698	−3.17

[a] From Eyring and Jhon (1969, p. 75).

table, the average cluster size decreases with an increase in temperature, while the mole fraction of unbonded water increases at the same time, all in accord with expectation. However, the fraction of nonbonded water is inconsistent with absorption spectroscopic experiments (Symons, 1972; Stevenson, 1965) in which the small amount of monomer water was found. Since a complete partition function has explicit dependence on volume, the molar volumes were computed by introducing three more adjustable parameters without explicit justification. In addition, there are difficulties in the partition function. (1) The combinational factor in the partition function leads to an overcount of the number of configurations. (2) The assumption that only one cluster size (besides the monomer) exists at a given temperature is an oversimplification. In spite of these defects, Neméthy and Scheraga's model has been widely accepted and used. In particular, their treatment has been helpful in understanding the nature of protein solutions and protein conformational stability.

2. *Modification of the Model by Hagler et al.*

Very recently, Hagler *et al.* (1973) modified the model of Neméthy and Scheraga and have corrected various deficiencies of the earlier model. In their model, they consider distributions of size of the cluster in addition to monomers, instead of a fixed size for each cluster. The internal energy of clusters is computed by assuming it to be the sum of the energies of the hydrogen bonds in the cluster. Again, the values of the hydrogen bond energy is one of the adjustable parameters in the model. On this basis, a modified partition function analogous to that of Neméthy and Scheraga was constructed. The distribution of cluster sizes, the median cluster size, and the mole fraction of unbroken hydrogen bonds are evaluated to give the best fit to the experimental thermodynamic data between 0° and 60°C. The agreement between theory and experiment is qualitatively good for the radial distribution, the volume–temperature behavior, and the infrared spectrum of water. According to the model, most of the clusters are found to be rather small, with the median of the distribution giving 11.2 at 0°C to 5.6 at 60°C. At no temperature is there a significant number of large clusters (> 60).

Approximations in the theory render it inaccurate for the treatment of very small clusters. For this reason, the theory is not applicable at temperatures above 60°C, where small clusters predominate. This is the difficulty that exists in the new model. Another approximation of the theory is the use of a single value for the H-bond energy regardless of the size and structure of the clusters. Work is in progress (Lentz *et al.*, 1974) to improve the model. Another problem is that some properties of clusters are assumed to vary with cluster size. An effort has been made to formulate estimates as to how these properties vary without introducing any new parameters (Lentz *et al.*, 1974).

C. Significant Structure Theory

Marchi and Eyring (1964) applied the significant structure theory to water. They assumed two solidlike species—a nonrotating hydrogen-bonded species and a rotating monomer—and calculated all thermodynamic properties with reasonably good results. However, the model does not exhibit a maximum in the density; the absorption-spectroscopic studies by Stevenson (1965) indicate that the rotating monomer concentration is very small. The small molar volume also makes it difficult to accept the substantial amount of rotating monomer the model envisions.

To overcome these difficulties, Jhon and Eyring and collaborators proposed a modified model and have developed it in a series of papers (Jhon *et al.*, 1966, 1967; Hobbs *et al.*, 1966; Jhon and Eyring, 1968b, 1976; Cho and Jhon, 1972; Bahng *et al.*, 1971) to explain thermodynamic, surface, dielectric, and transport properties. In addition, the model has been found to be useful in explaining the solubility of gases in water (Jhon *et al.*, 1971). The model is visualized as containing at least two solidlike clusters in equilibrium with each other and with fluidized vacancies conferring gaslike properties on part of the degrees of freedom. One of these structures is a cluster of about 46 molecules with a structure and density similar to ice I. The clusters of about 46 molecules are dispersed in an ice III-like structure which is 20% more dense than the ice I-like cluster which transforms cooperatively as a unit between the two states. Introducing these concepts, one obtains the partition function for water. The partition function for the solidlike part is as follows [see Eq. (2.1)].

$$(f_s f_{\text{deg}})^{N(V_s/V)} = (f_{s_1})^{n_1}(f_{s_3})^{n_3} = f_{s_3}^{N(V_s/V)} \cdot \left(\frac{f_{s_1}}{f_{s_3}}\right)^{n_1} = (f_{s_3} K^{(1+K^{-q})^{-1}})^{N(V_s/V)} \quad (6.4)$$

where f_{s_1} and f_{s_3} are the partition function for the ice I-like and ice III-like molecules, and n_1 and n_3 are the numbers of molecules of the respective species. We designate the fraction of ice I-like and ice III-like clusters by N_1 and N_3, respectively, where $N_1 + N_3 = 1$.

$$\frac{N_1}{N_3} = \frac{N_1}{1 - N_1} = \left(\frac{f_{s1}}{f_{s3}}\right)^q = K^q$$

so that

$$N_1 = \frac{K^q}{1 + K^q} = (1 + K^{-q})^{-1}$$

and $n_1 = NV_s/V(1 + K^{-q})^{-1}$. This value for N_1 explains the last equality in Eq. (6.4). Equation (6.4) can be rewritten in the usual form of significant structure theory.

$$f_s f_{\text{deg}} = \frac{\exp(E_s/RT)}{(1 - e^{-\theta/T})^6}\left[1 + n\left(\frac{V - V_s}{V_s}\right)\exp\left(-\frac{aE_s V_s}{(V - V_s)RT}\right)\right]$$
$$\times \prod_{i=1}^{3} \frac{1}{1 - \exp(-h\nu_i/kT)}(K)^{[1/(1+K^{-q})]}. \tag{6.5}$$

In deriving Eq. (6.5), we assume that the molecules in both ice I-like and ice III-like clusters do not rotate and that an equilibrium between them is established cooperatively.

q Ice I-like cluster $\rightleftarrows q$ Ice III-like cluster.

Here the equilibrium constant K^q and the solidlike volume V_s is given by

$$K^q = [\text{Ice I-like}]/[\text{Ice III-like}] = (f_{s1}/f_{s3})^q$$
$$= \left[\exp - \frac{\Delta H}{RT}\exp\frac{\Delta S}{R}\exp - \frac{p\Delta V}{RT}\right]^q \tag{6.6}$$

where $\Delta V = V_{s1} - V_{s3}$ and

$$V_s = \frac{K^q}{1 + K^q}V_{s1} + \frac{1}{1 + K^q}V_{s3} = \frac{K^q V_{s1} + V_{s3}}{1 + K^q}. \tag{6.7}$$

q is the number of molecules in clusters, V_{s_1} and V_{s3} are the molar volumes of ice I-like molecules, and ice III-like molecules, respectively. For the gaslike partition function, the following form is used:

$$f_g = \frac{(2\pi mkT)^{3/2}}{h^3}\frac{eV}{N}\frac{8\pi^2(8\pi^3 ABC)^{1/2}(kT)^{3/2}}{2h^3}$$
$$\prod_{i=1}^{6}[1 - \exp(-h\nu_i/kT)]^{-1}. \tag{6.8}$$

All the symbols in Eqs. (6.5) to (6.8) have their usual statistical mechanical significance. From Eq. (2.1) and introducing Eqs. (6.5) and (6.8), one obtains the following partition function:

$$f = \left\{ \frac{\exp(E_s/kT)}{(1 - \exp(-\theta/T)^6} \right.$$

$$\times \left[1 + n\left(\frac{V - V_s}{V_s}\right) \exp\left(-\frac{aE_s V_s}{(V - V_s)RT}\right)\right] (K)^{(1+K^{-q})^{-1}} \Bigg\}^{N(V_s/V)}$$

$$\times \left\{ \frac{(2\pi mkT)^{3/2}}{h^3} \frac{eV}{N} \frac{8\pi^2(8\pi^3 ABC)^{1/2}(kT)^{3/2}}{2h^3} \right\}^{N[(V - V_s)/V]}$$

$$\times \left\{ \prod_{i=1}^{3} \left[1 - \exp\left(-\frac{h\nu_i}{kT}\right)\right]^{-1} \right\}^N. \tag{6.9}$$

Among various models, this has been most widely applied to explain almost all thermodynamic properties of water with success. In particular, the model successfully predicts the maximum density at 4°C, and a minimum in the heat capacity versus temperature. Calculated results for liquid volume and vapor pressure are compared with experiment in Table XII. The agreement between theory and experiment for the second derivatives of the partition function such as C_p constitute a reliable and severe test of the model. These and other results are presented in the paper by Jhon *et al.* (1966). So far, the most extensive fit to the viscosity data for water as a function of both temperature and pressure is achieved using this model. The curve of viscosity versus pressure below 30°C shows first a peculiar decrease in viscosity with pressure increase at low pressures and then an increase with pressure. This fact can be easily explained from the model. Assuming that water is a solution of the clusters of ice I-like molecules of negligible fluidity in the solvent of ice III-like molecules, the viscosity of water is given by the Einstein (1906, 1911) equation

$$\eta = \eta_3(1 + 2.5\phi + \cdots). \tag{6.10}$$

Here η_3 is the viscosity of the ice III-like structure given by

$$\eta_3 = C \exp(\Delta H^{\ddagger}/RT) \exp(p\Delta V^{\ddagger}/RT) \tag{6.11}$$

and ϕ is the volume fraction of the nonfluid ice I-like clusters in the system, i.e.,

$$\phi = \left[\frac{1}{1 + K^{-q}}\right]\left[\frac{V_{s1}}{V}\right]. \tag{6.12}$$

The good agreement between theory and experiment is seen in Fig. 18. Since ϕ in Eq. (6.10) decreases with pressure, the viscosity decreases with

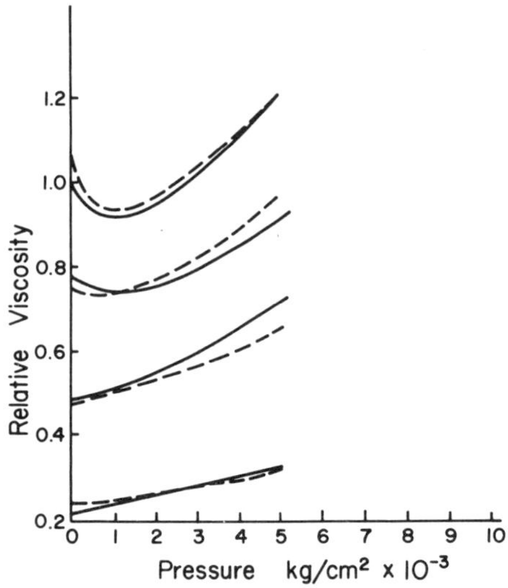

Fig. 18. The relative viscosity of water as a function of pressure. The curves are for the temperature 0°, 10.3°, 30°, and 75°C starting from the top. The relative viscosity, following Bridgman, relates all the viscosities to its value at 0°C and 1 atm. (——) experimental, (- - -) calculated.

pressure. On the other hand, because of the usual effect of pressure in squeezing out vacancies, the viscosity tends to increase with pressure [see Eq. (6.11)]. Thus, the appearance of the minimum in the curve of η vs. p is the net result of these two opposing effects. Another merit of this model is that the proposed structure for a cluster leads to the high observed dielectric constant of water. Pauling's (1959) cage structure predicts too small a value for the dielectric constant.

Although our model has been used to calculate many properties successfully, it involves some obvious simplifications. Six solidlike degrees of freedom were treated as three rotational and three translational Einstein oscillators having a single mean characteristic temperature θ. Inclusion of the explicit dependence of V_s on the temperature and pressure would be an improvement. While Neméthy, Scheraga and others assumed the cluster size changed with temperature, the cluster size here is assumed to be approximately 46 at any temperature with only the concentration changing.

To check the stability of the optimum cluster size of 46, the theory of intermolecular forces was introduced and found to provide some justification. Cho and Jhon (1972) calculated the optimum potential energy for such a cluster by means of the Lippincott–Schroeder empirical potential function for the hydrogen bond, the dispersion energy, and the dipole–dipole interaction terms. They found that an energetically most stable cluster consisted of approximately 46 molecules.

D. Modified Cell Theory of Weres and Rice

Weres and Rice (1971) developed a new modified cell model which they applied to water. In their model, they used the Ben-Naim and Stillinger intermolecular potential function (Ben-Naim, 1970; Ben-Naim and Stillinger, 1971). They introduced the following new effective pairwise additive intermolecular potential function:

$$V(X_1, X_2) = V_{LJ}(R_{12}) + S(R_{12})V_{HB}(X_1, X_2). \tag{6.13}$$

Here X_i $(i = 1, 2)$ are the generalized coordinates of molecule i specifying the position and orientation of that molecule. $V_{LJ}(R_{12})$ is a Lennard-Jones (6-12) potential with the neon atom parameters that contains the same number of electrons as the water molecule, and V_{HB} is a slight variant of the Bjerrun four-point charge model. The function $S\ (R_{12})$ is a switching function to give a small extra weight to the configuration in which the two point charges on neighboring molecules overlap.

The following clearly defined structure was assumed. In determining the most probable structure, the factors such as the short-range repulsive forces, the hydrogen bonding geometry, the high density of liquid water, and its highly negative internal energy are considered. These calculations led them to the conclusion that the most probable structure in water is a partially occupied ice VIII lattice that consists of two interpenetrating diamondlike lattices. The distribution of the molecules in a fully occupied ice VIII lattice form a body centered cubic lattice. For the fundamental unit (or "basic cell") size they chose a regular octahedron in the body-centered cubic lattice which contains a central site and its fourteen nearest and next nearest neighbor sites. With full ice VIII occupancy, there is a central molecule with its four H-bonded nearest neighbors forming a diamondlike sublattice with ten molecules from another sublattice (see Fig. 19). They also restricted the range of the basic cell configurations that they deal with to those that have two, three, or four hydrogen-bonded neighbors.

The existence of substantial amounts of nonhydrogen-bonded molecules seems precluded by the spectroscopic data of Stevenson (1965). As mentioned earlier, one of the basic defects of the cell model is that it is unable to include the correlation effects of the motion of neighboring molecules. The best way of calculating the communal effects is to calculate the full intermolecular frequency spectrum in the quasi-harmonic approximation and subtract out the spectrum of the uncorrelated motions of single molecules. In their calculations, the communal effects were considered in this way. In the calculation of the entropy, the conventional method is used to obtain the configurational entropy and an orientational entropy calculation is made, which is an extension of Pauling's calculation for ice (see Eisenberg and Kauzmann, 1969). Many thermodynamical properties and the equilibrium

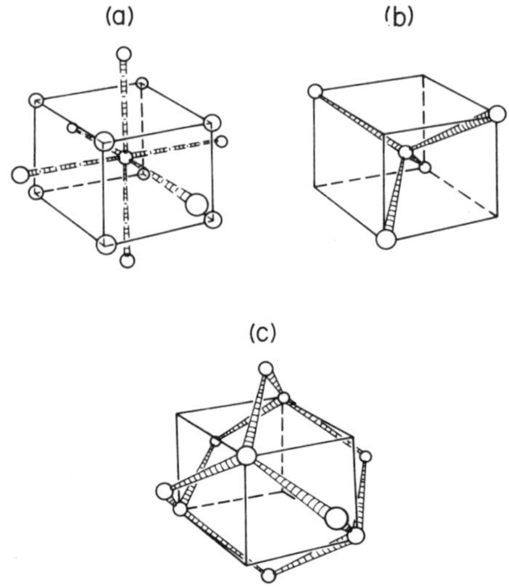

Fig. 19. Geometric characteristics of the present model: (a) space lattice section employed; (b) first molecular framework: (c) second molecular framework. Heavy lines indicate hydrogen bonds.

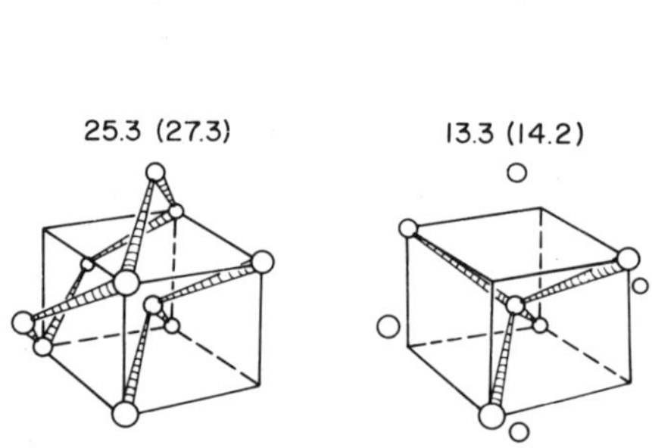

Fig. 20. The four most important basic cell configurations. The numbers above each represent its percentage concentration at $t = 0°$ and $t = 100°C$, the latter in parentheses.

basic cell concentrations were obtained by minimizing the free energy of the system.

According to their calculations, the Ben-Naim–Stillinger potential contains a serious error in its curvature dependence, which affects the values for the rotational degrees of freedom, and the quantum mechanical effects in water are too large to be neglected. The calculated results of the four most important basic cell configurations are shown in Fig. 20. The fractions of molecules with two, three, or four hydrogen-bonded neighbors are shown to be nearly constant throughout the temperature range studied. One of the major defects of the model is that the results it generates are difficult to compare with the experimental radial distribution function (Narten *et al.*, 1967; Narten and Levy, 1969). To develop the model, the use of an improved potential function using more than four charges considered in the Ben-Naim–Stillinger potential is one possible approach. As many other models do, this one also fails to predict a maximum density in water. Certain of the discrepancies between theory and experiment are qualitatively understood to arise from the well-known defects of the cell model.

E. Models of Aqueous and Ionic Solutions

Several calculations of the thermodynamic and bonding properties of aqueous hydrocarbon solutions have been carried out. Neméthy and Scheraga (1962, 1969) developed a model of solutions based on the Frank and Wen (1957) flickering cluster ideas using a modified form of the significant liquid structure approach. Herman (1971) has applied a modified significant structure theory of water to aqueous hydrocarbon solutions. Following this, Sung *et al.* (1972) developed a model for aqueous hydrocarbon solutions using their pure water model. The applications of the liquid model to ionic solution has not progressed far because of the problems of ionic solution arising from hydration, the mixing factor, and the electrostatic forces, etc. Recent applications of significant structure theory to ionic solutions has yielded encouraging results (Jhon and Eyring, 1968a). This model successfully explains many of the thermodynamic properties of ionic solutions. Subsequently, this model was applied to evaluate the hydration number for ions and the activity coefficient of the solute (Sung and Jhon, 1970a,b). We next consider various models in some detail.

1. The Neméthy and Scheraga Model for Aqueous Solutions of Hydrocarbons

The physical model for the structure of aqueous hydrocarbon solutions is based on their model for pure water (Neméthy and Scheraga, 1962, 1969). According to them, an essentially similar structure is assumed for water in

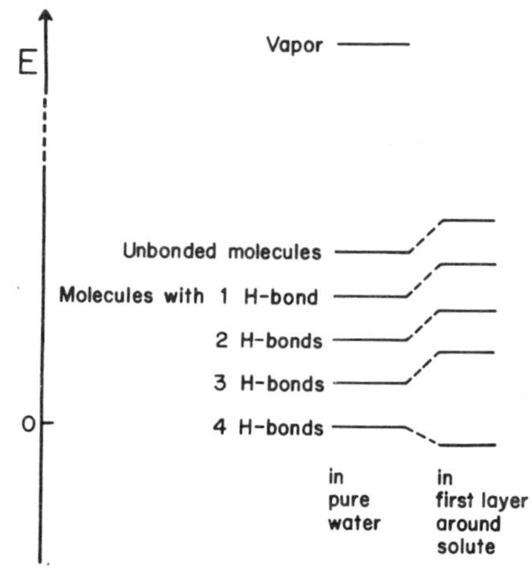

Fig. 21. Schematic representation of the energy level changes occurring on transfer of water from the pure liquid to the structure next to a nonpolar solute. It should be emphasized that these shifts in energy levels occur only for water molecules in the first layer around the hydrocarbon.

aqueous hydrocarbon solutions. The only difference for the water molecules is that the energy levels in the water layer next to the hydrocarbon are shifted due to different interactions between the water and hydrocarbon (see Fig. 21). It is understood that the energy shifts bring an increase in "ice-likeness" or an increase of hydration number near the hydrocarbon surface. The formation of a partial "ice cage" may occur around part of the hydrocarbon molecule similar to the gas hydrate structure. Then, one can write the partition function for a dilute solution Z_{sol} containing N_{A} moles of solute A and N_{w} mole of water as follows:

$$Z_{\text{sol}} = g(Z_{\text{A}})^{N_{\text{A}}N_0}(Z_{\text{w}}^c)^{Y^c N_{\text{A}} N_0}(Z_{\text{w}}^0)^{(N_2 - Y^c N_{\text{A}})N_0}. \tag{6.14}$$

Here Y^cN_{A} and $N_{\text{w}} - Y^cN_{\text{A}}$ indicate the moles of water in the first layer near the solute, and moles of unperturbed water, respectively. In the combinatorial factor, N_0 is Avogadro's number, and Z_{A}, Z_{w}^c Z_{w}^0 are the partition functions for the solute in water, the first layer water about the hydrocarbon, and the bulk water, respectively. It is understood that the partition function of pure solute G_{R} has only a different value of the interaction term compared to that of the solute in water Z_{A}, since the former involves the interaction between solute molecules. Utilizing the free energy $A = -kT \ln Z_{\text{sol}}$, and the above considerations, one obtains the Helmholtz free energy change of the solution process. In addition, the enthalpies and entropies of the solution are calculated for many hydrocarbons. However, in their calculations, the several extra adjustable parameters, such as the hydration number of the solute, the energy shifts for the various water species, and the interaction energy difference between solute pair and solute-water pair, are made to best fit the experimental data. The volume changes of solution are derived semiquantitatively.

2. *The Significant Structure Model of Hydrocarbons in Water*

Hermann has applied a modified form of the significant structure theory of water to the aqueous hydrocarbon solutions. Values for the properties treated by Neméthy and Scheraga are calculated. Unfortunately, there are some unclear physical concepts in the formulation of the partition function.

Recently, Sung *et al.* (1972) developed such a model by simply utilizing early published results in the following way: The partition functions of bulk water (Jhon *et al.*, 1966) as well as that of surface water molecules (Jhon *et al.*, 1967a) is known. In the aqueous hydrocarbon solutions, there are hydrocarbon molecules, bulk water molecules, and water molecules adjacent to the hydrocarbon molecules that behave like molecules at the air–water interface. Thus, the partition function for the solution of a hydrocarbon in water may be expressed as

$$Z_{\text{sol}} \equiv M Z_{\text{R}}^{N_2} Z_{\text{w}}^{N_1 - LN_2} Z_{\text{L}}^{LN_2} \tag{6.15}$$

where M is a combinatorial factor, Z_{R}, Z_{W}, and Z_{L} are the partition functions for the hydrocarbon, the bulk water, and adjacent water next to the

Table XIII

CALCULATED AND OBSERVED THERMODYNAMIC PROPERTIES OF HYDROCARBONS IN AQUEOUS SOLUTION AT 298.15°K[a]

	ΔE(kcal/mole)		ΔS(*e.u.*)		ΔC_V(cal/deg·mole)	
Hydrocarbons	Calc.	Obs.	Calc.	Obs.	Calc.	Ref.[b]
Methane	−2.04	−2.86 ~ −2.25	−15.3	−18.4 ~ −16.8	8.5	
Ethane	−1.88	−2.37 ~ −1.27	−18.7	−19.5 ~ −16.8	10.3	
Propane	−1.17	−2.09 ~ −1.45	−22.1	−23.5 ~ −21.3	12.2	7.1
Butane	−0.50	−0.95 ~ −0.72	−25.5	−22.7 ~ −21.9	14.1	8.3
Benzene	0.49	0.58	−13.6	−13.5	17.5	9.8
Toluene	0.58	0.64	−15.8	−15.7	18.8	11.0
m-Xylene	0.37	0.41	−19.2	−19.1	20.3	—
p-Xylene	0.39	0.46	−18.7	−19.0	20.2	—
Ethylbenzene	0.25	0.39	−19.0	−19.0	20.6	—
Isopropylbenzene	0.93	1.06	−19.7	−19.1	21.0	—

[a] From Sung *et al.* (1972).

[b] From Hermann (1971).

hydrocarbons with partition functions analogous to these for molecules at the air–water interface. Also, L is the number of water molecules adjacent to a hydrocarbon molecule, N is Avogadro's number, and N_1 and N_2 are the total numbers of water and hydrocarbon molecules, respectively. As mentioned previously, the partition function for pure hydrocarbon G_R is different from that of a hydrocarbon in water Z_R. Several thermodynamic properties such as the Helmholtz free energy, the internal energy, the entropy, and the heat capacity have been calculated. Some of them are shown in Table XIII. One sees good agreement between theory and experiment. This procedure is more economical of parameters than previous treatments have been.

3. *The Significant Structure Model Applied to Ionic Solutions*

The significant structure theory has been applied to ionic solutions. The results so far obtained are encouraging (Jhon and Eyring, 1968a). The partition function used is

$$f_{\text{ionic solution}} = \left\{\frac{\exp(E_{\text{smix}}/RT)}{(1 - e^{-\theta_1/T})^3(1 - e^{-\theta_2/T})^3}\right\}^{m'N}$$

$$\times (K_1)^{[m'n_1N/(1+K_1^{-1})]}(K_2)^{[m'n_2N/(1+K_2^{-1})]}(f_{\text{pure H}_2\text{O}})^{55.51N}$$

$$\times \exp\left\{\frac{V}{8\pi}\frac{1}{1 + a'\kappa}\left(\kappa^3 - \frac{1}{(a')^3}\right) + \frac{1}{(a')^3}(1 + a'\kappa)\right.$$

$$\left. - \frac{2}{(a')^3}\ln(1 + a'\kappa)\right\}\frac{(55.51N + 2m'N)!}{(55.51N)!\,(m'N)!)^2} \tag{6.16}$$

where m', K_1, K_2, θ_1, θ_2, N_1, N_2, E_{smix}, a' and κ are the molarity of the salt, the equilibrium constant between water absorbed on a cation site and free water, the corresponding equilibrium constants for the anion, the Einstein characteristic temperature for the cation, and for the anion, the number of sites on an anion and on a cation, the energy of sublimation of the solute, the ionic radius, and the Debye–Hückel parameter. In deriving Eq. (6.16), the partition function of hydrated water is replaced by the equilibrium constant K_i ($i = 1.2$) between free water and a hydrated molecule. And the electrostatic term due to the ions is derived from the Debye–Hückel theory. The additional contribution of the Helmholtz free energy is given in the fifth term on the right-hand side of Eq. (6.16). This model successfully explains many of the thermodynamic properties of ionic solutions. The model has been carried further to obtain the hydration number for ions and the activity coefficient of the solute (Sung *et al.*, 1970a,b).

References

Adams, D. J., and Matheson, A. J. (1972). *J. Chem. Soc., Faraday Trans. 2* **68**, 1536.
Adams, D. J., and Matheson, A. J. (1973). *Chem. Phys. Lett.* **22**, 484.
Ahn, W. S., Pak, H., and Chang, S. (1965). *J. Korean Chem. Soc.* **9**, 215.
Alder, B. J., and Wainwright, T. E. (1957). *J. Chem. Phys.* **27**, 1208.
Alder, B. J., and Wainwright, T. E. (1962). *Phys. Rev.* **127**, 359.
Alder, B. J., Hoover, W. G., and Wainwright, T. E. (1963). *Phys. Rev. Lett.* **11**, 241.
Andrade, E. N. da C. (1930). *Nature (London)* **125**, 309.
Bahng, J. S., Hahn, S. J., and Jhon, M. S. (1970). *J. Korean Chem. Soc.* **14**, 193.
Bahng, J. S., Hahn, S. J., and Jhon, M. S. (1971). *J. Korean Chem. Soc.* **15**, 171.
Barker, J. A. (1955). *Proc. R. Soc. London, Ser. A* **230**, 390.
Barker, J. A. (1956). *Proc. R. Soc. London, Ser. A* **237**, 63.
Barker, J. A. (1957). *Proc. R. Soc. London, Ser. A* **240**, 265.
Barker, J. A. (1958). *Proc. R. Soc. London, Ser. A* **241**, 547.
Barker, J. A. (1960). *Aust. J. Chem.* **13**, 187.
Barker, J. A. (1961). *Proc. R. Soc. London, Ser. A* **259**, 442.
Barker, J. A. (1962). *Aust. J. Phys.* **15**, 127.
Barker, J. A. (1963). "Lattice Theories of the Liquid State." Pergamon, Oxford.
Barker, J. A., and Henderson, D. (1968). *J. Chem. Educ.* **45**, 2.
Bell, G. M., and Lavis, D. A. (1970). *J. Phys. A: Gen. Phys.* **3**, 568.
Ben-Naim, A. (1970). *J. Chem. Phys.* **52**, 5531.
Ben-Naim, A., and Stillinger, F. H. (1971). *In* "Structure and Transport Processes in Water and Aqueous Solutions" (R. A. Horne, ed.), Chapter 8. Wiley (Interscience), New York.
Blömgren, G. E. (1960). *Ann. N.Y. Acad. Sci.* **79**, 781.
Buehler, R. J., Wentorf, R. H., Jr., Hirschfelder, J. O., and Curtiss, C. F. (1951). *J. Chem. Phys.* **19**, 61.
Caillelet, L., and Mathias, E. (1886). *C.R. Hebd. Seances Acad. Sci.* **102**, 1202.
Caillelet, L., and Mathias, E. (1887). *C.R. Hebd. Seances Acad. Sci.* **104**, 1563.
Callaway, J. (1959). *Phys. Rev.* **113**, 1046.
Carlson, C. M., Eyring, H., and Ree, T. (1960a). *Proc. Natl. Acad. Sci. U.S.A.* **46**, 333.
Carlson, C. M., Eyring, H., and Ree, T. (1960b). *Proc. Natl. Acad. Sci. U.S.A.* **46**, 649.
Cernuschi, F., and Eyring, H. (1939). *J. Chem. Phys.* **7**, 547.
Chang, S., Ree, T., Eyring, H., and Matzner, I. (1962). *In* "Progress in International Research on Thermodynamic and Transport Properties" (J. F. Masi and D. H. Tsai, eds.), p. 88. Academic Press, New York.
Cho, U., and Jhon, M. S. (1972). *J. Korean Chem. Soc.* **15**, 171.
Choi, S. H., Do, Y. K., and Jhon, M. S. (1977). To be published.
Chung, H. S., and Dahler, J. S. (1964). *J. Chem. Phys.* **40**, 2868.
Chung, H. S., and Dahler, J. S. (1965a). *J. Chem. Phys.* **42**, 2374.
Chung, H. S., and Dahler, J. S. (1965b). *J. Chem. Phys.* **43**, 2306.
Cohen, E. G. D., and Rethmeier, B. C. (1958). *Physica (Utrecht)* **24**, 959.
Cohen, E. G. D., de Boer, J., and Salsburg, Z. W. (1955). *Physica (Utrecht)* **21**, 137.
Cohen, M. H., and Turnbull, D. (1959). *J. Chem. Phys.* **31**, 1164.
Dahler, J. S., and Cohen, E. G. D. (1966). *Physica (Utrecht)* **26**, 81.
Dahler, J. S., Hirschfelder, J. O., and Thacher, H. C. (1956). *J. Chem. Phys.* **25**, 249.
David, H. G., and Hamann, S. D. (1953). *Trans. Faraday Soc.* **49**, 711.
de Boer, J. (1954). *Physica (Utrecht)* **20**, 655.
de Boer, J., and Lunbeck, R. J. (1948). *Physica (Utrecht)* **14**, 320.

Defay, R., Prigogine, I., Bellemans, A., and Everett, D. H. (1966). "Surface Tension and Adsorption." Wiley, New York.

Devonshire, A. F. (1937). *Proc. R. Soc. London, Ser. A* **163**, 132.

Do, Y. K., Jhon, M. S., and Ree, T. (1976). *J. Korean Chem. Soc.* **20**, 118.

Doolittle, A. K. (1951). *J. Appl. Phys.* **22**, 1471.

Doolittle, A. K. (1952). *J. Appl. Phys.* **23**, 236.

Eckert, C. A., and Prausnitz, J. M. (1964). *AIChEJ* **10**, 677.

Egelstaff, P. A. (1967). "An Introduction to the Liquid State." Academic Press, New York.

Einstein, A. (1906). *Ann. Phys. (Leipzig)* [4] **19**, 289.

Einstein, A. (1911). *Ann. Phys. (Leipzig)* [4] **34**, 591.

Eisenberg, D., and Kauzmann, W. (1969). "The Structure and Properties of Water." Oxford Univ. Press, London and New York.

Eisenstein, A., and Gingrich, N. (1942). *Phys. Rev.* **62**, 261.

Englert-Chowles, A. (1955). *J. Chem. Phys.* **23**, 1168.

Englert-Chowles, A., and Prigogine, I. (1958a). *J. Chim. Phys.* **55**, 16.

Englert-Chowles, A., and Prigogine, I. (1958b). *Nuovo Cimento, Suppl.* **9**, 347.

Eyring, H. (1936). *J. Chem. Phys.* **4**, 283.

Eyring, H., and Hirschfelder, J. O. (1937). *J. Phys. Chem.* **41**, 249.

Eyring, H., and Jhon, M. S. (1969). "Significant Liquid Structures." Wiley, New York.

Eyring, H., and Ree, T. (1961). *Proc. Natl. Acad. Sci. U.S.A.* **47**, 526.

Eyring, H., Ree, T., and Hirai, N. (1958a). *Proc. Natl. Acad. Sci. U.S.A.* **44**, 683.

Eyring, H., Ree, T., and Hirai, N. (1958b). *Proc. Natl. Acad. Sci. U.S.A.* **44**, 1213.

Eyring, H., Henderson, D., and Jost, W., eds. (1971). "Physical Chemistry—An Advanced Treatise." Academic Press, New York.

Faerber, G. L., Kim, S. W., and Eyring, H. (1970). *J. Phys. Chem.* **74**, 3510.

Faerber, G. L., Breitling, S. M., MacKnight, A., and Eyring, H. (1972). *J. Phys. Chem.* **76**, 731.

Flory, P. J. (1953). "Principles of Polymer Chemistry." Cornell Univ. Press, Ithaca, New York.

Frank, H. S., and Wen, W. Y. (1957). *Discuss. Faraday Soc.* **24**, 133.

Franks, F. F., ed. (1972). "Water—A Comprehensive Treatise," Vols. 1–4. Plenum, New York.

Frenkel, J. (1955). "Kinetic Theory of Liquids." Dover, New York.

Fuller, E. J., Ree, T., and Eyring, H. (1959). *Proc. Natl. Acad. Sci. U.S.A.* **45**, 1594.

Glasstone, S. N., Laidler, K., and Eyring, H. (1941). "The Theory of Rate Processes." McGraw-Hill, New York.

Grindlay, J. (1961). *Proc. Phys. Soc. London* **77**, 1001.

Grosh, J., Jhon, M. S., Ree, T., and Eyring, H. (1967). *Proc. Natl. Acad. Sci. U.S.A.* **57**, 1566.

Guggenheim, E. A. (1952). "Mixtures." Oxford Univ. Press, London and New York.

Hagler, A. T., Scheraga, H. A., and Neméthy, G. (1973). *Ann. N.Y. Acad. Sci.* **204**, 51.

Hamann, S. D. (1952). *Trans. Faraday Soc.* **48**, 303.

Hamann, S. D. (1954). *J. Am. Chem. Soc.* **76**, 4244.

Henderson, D. (1962). *J. Chem. Phys.* **37**, 631.

Henderson, D. (1963). *J. Chem. Phys.* **39**, 1887.

Henderson, D. (1974). *Annu. Rev. Phys. Chem.* **25**, 461.

Henderson, D., and Reed, R. D. (1964). *J. Chem. Phys.* **40**, 975.

Henderson, D., Eyring, H., and Felix, D. (1962). *J. Phys. Chem.* **66**, 1128.

Herman, R. B. (1971). *J. Phys. Chem.* **75**, 363.

Hildebrand, J. H., and Scott, R. L. (1950). "Solubility of Nonelectrolytes," 3rd ed. Van Nostrand-Reinhold, Princeton, New Jersey.

Hildebrand, J. H., and Scott, R. L. (1962). "Regular Solutions." Prentice-Hall, Englewood Cliffs, New Jersey.

Hill, T. L. (1947). *J. Phys. Colloid Chem.* **51**, 1219.

Hillier, I., and Walkley, J. (1964). *J. Chem. Phys.* **41**, 2168.

Hirschfelder, J. O., Stevenson, D., and Eyring, H. (1937). *J. Chem. Phys.* **5**, 896.

Hirschfelder, J. O., Curtiss, C. F., and Bird, R. B. (1954). "Molecular Theory of Gases and Liquids." Wiley, New York.

Hobbs, M. E., Jhon, M. S., and Eyring, H. (1966). *Proc. Natl. Acad. Sci. U.S.A.* **56**, 31.

Hoory, S. E., and Prausnitz, J. M. (1967). *Trans. Faraday Soc.* **63**, 455.

Horne, R. A., ed. (1972). "Water and Aqueous Solutions." Wiley (Interscience), New York.

Hsu, C. C., and Eyring, H. (1972a). *Proc. Natl. Acad. Sci. U.S.A.* **69**, 1125.

Hsu, C. C., and Eyring, H. (1972b). *Proc. Natl. Acad. Sci. U.S.A.* **69**, 1342.

Hsu, C. C., MacKnight, A. K., and Eyring, H. (1972). *J. Phys. Chem.* **76**, 1012.

Huggins, M. L. (1962). "Physical Chemistry of High Polymers." Wiley, New York.

Huggins, M. L. (1970). *J. Phys. Chem.* **74**, 371.

Jancso, G., and van Hook, W. A. (1974). *Chem. Rev.* **74**, 689.

Janssens, P., and Prigogine, I. (1950). *Physica (Utrecht)* **16**, 895.

Jhon, M. S. (1967). *J. Korean Chem. Soc.* **11**, 60.

Jhon, M. S., and Eyring, H. (1968a). *ACS Southeast. Reg. Meet., 1968.*

Jhon, M. S., and Eyring, H. (1968b). *J. Am. Chem. Soc.* **90**, 3071.

Jhon, M. S., and Eyring, H. (1971). *Adv. Phys. Chem.* **8**, 335.

Jhon, M. S., Grosh, J., Ree, T., and Eyring, H. (1965). *Proc. Natl. Acad. Sci. U.S.A.* **54**, 1419.

Jhon, M. S., Grosh, J., Ree, T., and Eyring, H. (1966). *J. Chem. Phys.* **44**, 1465.

Jhon, M. S., van Artsdalen, E. R., Grosh, J., and Eyring, H. (1967a). *J. Chem. Phys.* **47**, 223.

Jhon, M. S., Grosh, J., and Eyring, H. (1967b). *J. Phys. Chem.* **71**, 2253.

Jhon, M. S., Clemena, G., and van Artsdalen, E. R. (1968). *J. Phys. Chem.* **72**, 4155.

Jhon, M. S., Klotz, W. L., and Eyring, H. (1969). *J. Chem. Phys.* **51**, 3692.

Jhon, M. S., Sung, Y. K., and Eyring, H. (1971). *Chem. Phys. Lett.* **13**, 36.

Kavanau, J. L. (1964). "Water and Solute-Water Interactions." Holden-Day, San Francisco, California.

Kiemens, P. G. (1951). *Proc. R. Soc. London, Ser. A* **208**, 108.

Kim, K. C., Lu, W. C., Ree, T., and Eyring, H. (1967). *Proc. Natl. Acad. Sci. U.S.A.* **57**, 861.

Kim, S. W., Jhon, M. S., Ree, T., and Eyring, H. (1968). *Proc. Natl. Acad. Sci. U.S.A.* **59**, 336.

Kim, S. W., Eyring, H., and Lee, Y. T. (1969). *J. Chem. Phys.* **51**, 3967.

Kim, U. S., Schmidt, R., Jhon, M. S., and Eyring, H. (1972). *Proc. Natl. Acad. Sci. U.S.A.* **69**, 1690.

Kim, U. S., Sung, Y. K., Jhon, M. S., and Eyring, H. (1976). *Phys. Chem. Liq.* **1**, 61.

Kirkwood, J. G. (1935). *J. Chem. Phys.* **3**, 300.

Kirkwood, J. G. (1950). *J. Chem. Phys.* **18**, 380.

Kirkwood, J. G., Lewinson, V. A., and Alder, B. J. (1952). *J. Chem. Phys.* **20**, 929.

Lee, H. B., Jhon, M. S., and Chang, S. (1964). *J. Korean Chem. Soc.* **8**, 179.

Lee, S. H., Jhon, M. S., and Eyring, H. (1977). *Proc. Natl. Acad. Sci. U.S.A.* **74**, 10.

Lennard-Jones, J. E., and Devonshire, A. F. (1937). *Proc. R. Soc. London, Ser. A* **163**, 53.

Lennard-Jones, J. E., and Devonshire, A. F. (1938). *Proc. R. Soc. London, Ser. A* **165**, 1.

Lentz, B., Hagler, A. T., and Scheraga, H. A. (1974). *J. Phys. Chem.* **78**, 1531.

Levelt, J. M. H., and Hurst, R. P. (1960). *J. Chem. Phys.* **32**, 96.

Liang, K., Eyring, H., and Marchi, R. P. (1964). *Proc. Natl. Acad. Sci. U.S.A.* **52**, 1107.

Lin, S. H., Eyring, H., and Davis, W. J. (1964). *J. Phys. Chem.* **68**, 3021.

Loeb, L. B. (1927). "Kinetic Theory of Gases." McGraw-Hill, New York.

Lu, W. C., Jhon, M. S., Ree, T., and Eyring, H. (1967). *J. Chem. Phys.* **46**, 1075.

Lu, W. C., Ree, T., Gerrard, V. G., and Eyring, H. (1968). *J. Chem. Phys.* **49**, 797.

Lumsden, J. (1966). "Thermodynamics of Molten Salt Mixtures." Academic Press, New York.

Ma, S. M., and Eyring, H. (1965). *J. Chem. Phys.* **42**, 1920.

Ma, S. M., and Eyring, H. (1975). *Proc. Natl. Acad. Sci. U.S.A.* **72**, 78.
Ma, S., Eyring, H., and Jhon, M. S. (1974). *Proc. Natl. Acad. Sci. U.S.A.* **71**, 3096.
Ma, S. M., Lee, S. T., Yoon, Y. H., Liu, S. H., and Eyring, H. (1977). To be published.
McAlpin, J. J. (1963). Ph.D. Thesis, Georgia Inst. Tech., Atlanta.
McAlpin, J. J., and Pierotti, R. A. (1964). *J. Chem. Phys.* **41**, 68.
Macedo, P. B., and Litovitz, T. A. (1965). *J. Chem. Phys.* **42**, 245.
McLaughlin, D. R., and Eyring, H. (1966). *Proc. Natl. Acad. Sci. U.S.A.* **55**, 1031.
McQuarrie, D. A. (1962). *J. Phys. Chem.* **66**, 1508.
Malijevský, A. (1971). *Collect. Czech. Chem. Commun.* **36**, 2137.
Marchi, R. P., and Eyring, H. (1964). *J. Phys. Chem.* **68**, 221.
Matheson, A. A. (1966). *J. Chem. Phys.* **44**, 695.
Mathot, V., and Desmyter, A. (1953). *J. Chem. Phys.* **21**, 782.
Mayer, J. E., and Careri, G. (1952). *J. Chem. Phys.* **20**, 1001.
Miller, A. A. (1964). *J. Polym. Sci., Part A* **2**, 1095.
Miner, B. A., and Eyring, H. (1965). *Proc. Natl. Acad. Sci. U.S.A.* **53**, 1227.
Moore, W. J. (1962). "Physical Chemistry," 4th ed. Prentice-Hall, Englewood Cliffs, New Jersey.
Mullin, W. J. (1971). *Phys. Rev. A* **4**, 1247.
Nagata, I., and Nishida, R. (1969). *Kanazawa Daigaku Kogakubu Kiyo* **5**, 93.
Narten, A. H., and Levy, H. A. (1969). *Science* **165**, 447.
Narten, A. H., Anford, M. D., and Levy, H. A. (1967). *Discuss. Faraday Soc.* **43**, 97.
Neméthy, G., and Scheraga, H. A. (1962). *J. Chem. Phys.* **36**, 3382.
Neméthy, G., and Scheraga, H. A. (1969). *J. Chem. Phys.* **41**, 680.
Newsome, D. S. (1974). *J. Phys. Chem.* **78**, 2600.
Nissen, D. A., and Carlsten, R. W. (1974). *J. Electrochem. Soc.* **121**, 500.
Oden, L., and Henderson, D. (1964). *J. Chem. Phys.* **41**, 3487.
Ono, S. (1947). *Mem. Fac. Eng., Kyushu Univ.* **16**, 190.
Pace, E. L. (1957). *J. Chem. Phys.* **27**, 1341.
Pack, W. H., Sung, Y. K., and Jhon, M. S. (1970). *J. Korean Chem. Soc.* **14**, 263.
Pauling, L. (1959). *In* "Hydrogen Bonding" (L. Hadzi, ed.), p. 1. Pergamon, Oxford.
Peek, H. M., and Hill, T. L. (1950). *J. Chem. Phys.* **19**, 1519.
Peierls, R. (1929). *Ann. Phys. (Leipzig)* [5] **3**, 1055.
Pekeris, C. L. (1950). *Phys. Rev.* **79**, 884.
Percus, J. K., and Yevick, G. J. (1958). *Phys. Rev.* **110**, 1.
Pierotti, R. A. (1962). *J. Chem. Phys.* **36**, 2515.
Pierotti, R. A. (1965). *J. Chem. Phys.* **43**, 1072.
Pierotti, R. A., and Thomas, H. E. (1971). *Surf. Colloid Sci.* **4**, 93.
Pople, J. A. (1951). *Philos. Mag.* [7] **42**, 459.
Prenzlow, C. F., and Halsey, G. D. (1957). *J. Phys. Chem.* **61**, 1158.
Prenzlow, C. F., and Halsey, H. A. (see Prenzlow, C. F., Thesis, University of Washington, Seattle. Ann Arbor, Michigan, Publication 20395. 94 pp.)
Prigogine, I. (1957). "The Molecular Theory of Solutions." North-Holland Publ. Amsterdam.
Prigogine, I., and Garikian, C. T. (1948). *J. Chim. Phys.* **45**, 273.
Prigogine, I., and Mathot, V. (1952). *J. Chem. Phys.* **20**, 49.
Prigogine, I., and Raulier, S. (1942). *Physica (Utrecht)* **9**, 396.
Prigogine, I., and Saraga, L. (1952). *J. Chim. Phys.* **49**, 399.
Prigogine, I., Bellemans, A., and Englert-Chowles, A. (1956). *J. Chem. Phys.* **24**, 518.
Ree, T. S., Ree, T., and Eyring, H. (1962). *Proc. Natl. Acad. Sci. U.S.A.* **48**, 501.
Ree, T. S., Ree, T., and Eyring, H. (1964a). *Proc. Natl. Acad. Sci. U.S.A.* **51**, 344.
Ree, T. S., Ree, T., and Eyring, H. (1964b). *J. Phys. Chem.* **68**, 3262.

Ree, T. S., Ree, T., and Eyring, H. (1964c). *J. Chem. Phys.* **41**, 524.

Ree, T. S., Ree, T., Eyring, H., and Perkins, R. (1965). *J. Phys. Chem.* **69**, 3322.

Reed, R. D., and Henderson, D. (1964). *J. Chem. Phys.* **41**, 2705.

Ross, S., and Oliver, J. P. (1961). *J. Phys. Chem.* **65**, 608.

Rowlinson, J. S. (1959). "Liquids and Liquid Mixtures." Butterworth, London.

Rowlinson, J. S., and Curtiss, C. F. (1951). *J. Chem. Phys.* **19**, 1519.

Safford, G. J., Leung, P. S., Naumann, A. W., and Schaffer, P. C. (1969). *J. Chem. Phys.* **50**, 4444.

Salsburg, Z. W., and Kirkwood, J. G. (1952). *J. Chem. Phys.* **20**, 1538.

Salsburg, Z. W., and Kirkwood, J. G. (1953). *J. Chem. Phys.* **21**, 2169.

Salsburg, Z. W., Cohen, E. G. D., Rethmeier, B. C., and de Boer, J. (1957). *Physica (Utrecht)* **23**, 407.

Samoilov, O. Ya. (1965). "Structure of Aqueous—Electrolyte Solutions and Hydration of Ions." Consultants Bureau, New York.

Sams, J. R., Constabaris, G., and Halsey, G. D. (1962). *J. Phys. Chem.* **66**, 2154.

Senior, W. A., and Verall, R. E. (1972). *J. Phys. Chem.* **73**, 4242.

Sinaňoglu, O. (1974). *In* "The World of Quantum Chemistry" (R. Daudel and B. Pullman, eds.), p. 265. Reidel Publ., Dordrecht, Netherlands.

Sinaňoglu, O., and Pitzer, K. S. (1960). *J. Chem. Phys.* **32**, 1279.

Stevenson, D. P. (1965). *J. Phys. Chem.* **69**, 2145.

Stewart, C. W., and von Frankenberg, C. A. (1967). *J. Polym. Sci., Part A-2* **5**, 623.

Stewart, C. W., and von Frankenberg, C. A. (1968). *J. Polym. Sci., Part A-2* **6**, 1686.

Sung, Y. K., and Jhon, M. S. (1970a). *J. Korean Chem. Soc.* **14**, 185.

Sung, Y. K., and Jhon, M. S. (1970b) *J. Oceanogr. Soc. Korea* **5**, 1.

Sung, Y. K., Kim, U. S., and Jhon, M. S. (1972). *J. Korean Chem. Soc.* **16**, 265.

Symons, M. C. R. (1972). *Nature (London)* **239**, 257.

Turnbull, D., and Cohen, M. H. (1970). *J. Chem. Phys.* **52**, 3038.

Vilen, R., and Misdolea, C. (1966). *J. Chem. Phys.* **45**, 3414.

Vilen, R., and Misdolea, C. (1968). *J. Chem. Phys.* **49**, 3179.

Vinogradov, S. N., and Linnell, R. H. (1971). "Hydrogen Bonding." Van Nostrand-Reinhold, Princeton, New Jersey.

Wainwright, T. E., and Alder, B. J. (1958). *Nuovo Cimento, Suppl.* **9**, 116.

Wall, T. T., and Hornig, D. F. (1965). *J. Chem. Phys.* **43**, 2079.

Walrafen, G. E. (1968). *J. Chem. Phys.* **48**, 244.

Walter, J., and Eyring, H. (1941). *J. Chem. Phys.* **9**, 393.

Wang, Y. L., Ree, T., Ree, T. S., and Eyring, H. (1965). *J. Chem. Phys.* **42**, 1926.

Wang, Y. L., Ree, T., and Eyring, H. (1971). *J. Korean Chem. Soc.* **15**, 265 .

Weissmann, M. (1964). *J. Chem. Soc.* **40**, 175.

Weissmann, M., and Mazo, R. M. (1962). *J. Chem. Phys.* **37**, 2930.

Wentorf, R. H., Jr., Buehler, R. J., Hirschfelder, J. O., and Curtiss, C. F. (1950). *J. Chem. Phys.* **18**, 1484.

Weres, O., and Rice, S. A. (1971). *J. Am. Chem. Soc.* **94**, 8983.

Williams, M. L., Laudel, R. F., and Ferry, J. D. (1975). *J. Am. Chem. Soc.* **77**, 3701.

Wood, W. W. (1968). *In* "Physics of Simple Liquids," (H. N. V. Temperley, J. S. Rowlinson, and G. S. Rushbrooke, eds.), Chapter 5. North-Holland Publ., Amsterdam.

Wood, W. W., and Jacobsen, J. P. (1957). *J. Chem. Phys.* **27**, 1207.

Wu, S. M. L., MacKnight, A. K., Stearns, W. J., Jr., Zandler, M. E., and Eyring, H. (1972). *Phys. Chem. Liq.* **3**, 181.

Wunderlich, B. (1962). *J. Chem. Phys.* **37**, 1203.

Young, D. M., and Crowell, A. D. (1962). "Physical Adsorption of Gases." Butterworth, London.

Structure of Fluid ^{4}He

R. D. Murphy
Department of Physics
University of Missouri–Kansas City, Kansas City, Missouri

I. Introduction

It would appear at first sight that liquid helium should be one of the simplest of all liquids, since its high-temperature gaseous phase is one of the simplest of all gases. This behavior in the gaseous phase is a consequence of the fact that the interatomic forces in helium are quite simple: they are weak and central. In fact, however, the behavior of liquid helium is by no means simple: the two isotopes ^{3}He and ^{4}He form two quite different systems with very complex physical properties. This complexity is a consequence of quantum mechanical effects, and it is for this reason that the fluid helium systems are of great interest in themselves.

But there are other at least equally compelling reasons for the interest in the study of the properties of liquid helium, and the importance of this system transcends even the fascinating superfluid phenomena that occur. Complex as the helium liquids are—particularly the superfluid phases of liquid ^{3}He discovered at temperatures of the order of a millidegree and discussed in the review articles by Leggett (1975) and Wheatley (1975)—in fact the remark that the interactions between helium atoms is simple remains true. If theory is unable to account quantitatively for the properties of these systems, there is little reason to hope that it will be any more successful with other many-body systems, especially nuclei, where the interactions are considerably more difficult. Further, it is reasonable to expect that the methods that prove useful in studying the properties of liquid helium may be useful in other systems, and vice versa.

This article is not intended to be a comprehensive review of the physics of liquid helium, for which the reader is referred to the books of Atkins (1959), Wilks (1967), Keller (1969), and Putterman (1974). It will deal with a far narrower question, namely the question of the first-principles theoretical calculation of the structure of liquid ^{4}He. Even in that case, the intention here is not to give a comprehensive review of such calculations, as various aspects of this question have been discussed in review articles by Feenberg (1970), ter Haar (1971), Barker and Henderson (1976), and Campbell (1977). Rather, this article will be confined to discussing two different recent approaches to the problem. Both of these approaches involve a factorization of a many-body function into a product of pair functions; in both cases, this factorization reduces the problem to that of a fictitious interacting classical gas; and both methods appear reasonably promising and give good numerical results, a partial table of which will be presented here. One of the approaches (the Jastrow wave function) is valid at zero temperature; the other approach (the effective potential approximation) is valid at nonzero temperatures, even down to the critical point.

The plan of this paper is as follows. Section II will be devoted to a formulation of the problem, introduction of the appropriate notation, and discussion of the two-body potential used. Section III will contain a discussion of the zero-temperature results. Section IV will be devoted to the finite-temperature results, and in Section V will appear a discussion of the results and of the connection between the two methods.

II. Formulation of the Problem

We wish to consider in the canonical ensemble a system composed of N ^{4}He atoms of mass m confined to a volume V at a temperature T. We will denote by ρ the number density ($\rho \equiv N/V$) and will let $\beta = (kT)^{-1}$, where k is the Boltzmann constant. Then the Hamiltonian for such a system may be written

$$\hat{H} = \hat{K} + U_2 + U_3 + \cdots, \tag{1}$$

where

$$\hat{K} = -\sum_{i=1}^{N} \frac{\hbar^2}{2m} \nabla_i^2$$

$$U_2 = \sum_{i<j} u_2(r_{ij})$$

$$U_3 = \sum_{i<j<k} u_3(\mathbf{r}_i, \mathbf{r}_j, \mathbf{r}_k) \text{ etc.},$$

and $r_{ij} = |\mathbf{r}_i - \mathbf{r}_j|$. Here the quantity $\hat{K}$ is of course the operator for the kinetic energy, and the quantities U_n are the various contributions to the potential energy of interaction. The term u_2 represents the pairwise-additive potential, the term u_3 is the nonadditive three-body interaction (i.e., the potential energy of a system of three atoms that cannot be accounted for as the sum of pair interactions u_2), etc.

Turning first to the pair interaction, one notes that the fact that the atomic electrons in helium are tightly bound in 1s states means that the atoms have no permanent moments (i.e., they are spherically symmetric) and that the atomic polarizability of helium is small. Accordingly, one would expect the atomic forces to be rather weak (compared to other rare gas systems) and central. A number of forms of the two-body potential have been proposed, the best-known and most widely used of which is certainly the Lennard-Jones 12-6 potential,

$$u(r) = 4\varepsilon[(\sigma/r)^{12} - (\sigma/r)^6],$$

with the deBoer Michels choice of the two parameters ε and σ, namely $\varepsilon/k = 10.22$ K and $\sigma = 2.556$ Å. Although the asymptotic form of this potential $(-r^{-6})$ is correct, the repulsion (r^{-12}) is almost certainly too steep. There are alternate forms that are better founded. Notable among these are the semiempirical forms of Bruch and McGee (1967, 1970), especially the form that they call MDD-2, the parameters for which were obtained from virial coefficient and transport data. Another promising purely theoretical alternative is to be found in the preliminary calculations of Liu and McLean (1974), who have performed large-scale configuration–interaction calculations. In addition, there are purely experimental forms of the potential obtained from molecular beam scattering data by Farrar and Lee (1972), Siska *et al.* (1971), and Burgmans *et al.* (1976).

Although it is clear that the traditional Lennard-Jones potential is not correct, there is yet no consensus as to what potential would be better. This is unfortunate, in that there is evidence (Hansen, 1971; Murphy and McGee, 1973) that condensed-matter calculations are in fact rather sensitive to the pair potential. The procedure that seems most sensible in the face of the fact that so many calculations have been done for the Lennard-Jones potential is to do initial calculations for the Lennard-Jones potential in order to facilitate comparisons with other calculational schemes and then, where possible, to use better forms of the interaction. For this reason, the results presented here will be confined to the use of the Lennard-Jones potential, with results using more realistic forms of the two-body potential to be published elsewhere.

The fact, noted above, that the atomic polarizability of helium is small suggests strongly that the nonadditive three-body forces will be quite small. In the other rare gases, the largest contribution to the three-body energy comes from the Axilrod–Teller–Muto triple–dipole interaction (Axilrod and Teller, 1943; Muto, 1943); indeed in argon, owing to a possibly fortuitous cancellation which may or may not occur in helium, it appears to be sufficient to neglect all three-body terms except the triple–dipole term (Barker and Henderson, 1976). Explicit calculations (Hansen, 1970; Murphy and Barker, 1971) of the magnitude of the triple–dipole term in liquid ^{4}He show that it is about 2% of the binding energy of liquid helium. Since it is so small, it appears reasonable to treat it simply as a perturbation and, until other three-body terms like the dipole–quadrupole, etc., terms have been calculated explicitly, to ignore them. As to terms in U_4, U_5, etc., there is no evidence that they are of any consequence, and it seems entirely safe to neglect them.

III. The Zero-Temperature Approach

Even if we neglect all potential energy terms U_n for $n > 2$ in the Hamiltonian [Eq. (1)], it would be a hopeless task to solve the Schrödinger equation $H\psi = E_n\psi$, for the energy eigenvalues E_n or for the wave function ψ. Since an exact calculation in the sense of solving the Schrödinger equation is not possible, one looks to approximate methods. An obvious first guess would be to do a Hartree–Fock calculation, namely, to take

$$\psi(\mathbf{r}_1, \mathbf{r}_2, \ldots, \mathbf{r}_N) = \prod_{i=1}^{N} \phi(\mathbf{r}_i).$$

This form of the single-particle wave function ϕ would satisfy the symmetry requirement of the Bose–Einstein statistics obeyed by the ^{4}He atoms and ϕ could be determined variationally by requiring that it minimize the ground-state expectation value of H. Unfortunately, this procedure will not work; $\phi(r)$ turns out to be a plane wave and for the Lennard-Jones (or any other potential that does not possess a Fourier transform) the calculated ground-state energy is infinite. The correlations between the atoms simply cannot be neglected.

Clearly, the next guess would be to write

$$\psi(\mathbf{r}_1, \mathbf{r}_2, \ldots, \mathbf{r}_N) = \prod_{1 \le i < j \le N} f(r_{ij}) \tag{2}$$

and this form has come to be called the "Jastrow" wave function (Jastrow, 1955), although some authors more correctly call it the Bijl–Dingle–Jastrow wave function in recognition of earlier work (Bijl, 1940; Dingle, 1949). Again, this form satisfies the Bose–Einstein statistics. It is clear that Eq. (2) is simply the first term of a systematic expansion in which one writes

$$\psi = \prod_{i<j} f(r_{ij}) \prod_{i<j<k} h(r_{ij}, r_{ik}, r_{jk}) \cdots. \tag{3}$$

More discussion of this approximation and of its utility will be given in Section V.

If the ground-state wave function $\psi(\mathbf{r}_1, \mathbf{r}_2, \ldots, \mathbf{r}_N)$ is of the Jastrow form, then it is not very difficult to show that the ground-state expectation value of the Hamiltonian is given exactly by

$$\langle H \rangle = \langle \tilde{U}_2 \rangle + \langle U_3 \rangle + \cdots \tag{4}$$

where

$$\langle \tilde{U}_2 \rangle = \frac{1}{2} N\rho \int d^3r g_2(r) \left[u_2(r) - \frac{\hbar^2}{2m} \nabla^2 \ln f(r) \right]$$

$$\langle U_3 \rangle = \frac{1}{6} N\rho^2 \int d^3r_{12}\, d^3r_{23} g_3(\mathbf{r}_1, \mathbf{r}_2, \mathbf{r}_3) u_3(\mathbf{r}_1, \mathbf{r}_2, \mathbf{r}_3)$$

$$\vdots$$

The quantities $g_2, g_3, \ldots$ are the two-particle, three-particle, ... distribution functions, defined by

$$g_2(r_{12}) = N(N-1)\rho^{-2} \int d\mathbf{r}_3\, d\mathbf{r}_4 \cdots d\mathbf{r}_N \psi^2 \Big/ \int d\mathbf{r}_1 \cdots d\mathbf{r}_N \psi^2 \tag{5}$$

$$g_3(r_{12}, r_{13}, r_{23}) = N(N-1)(N-2)\rho^{-3} \int d\mathbf{r}_4\, d\mathbf{r}_5 \cdots d\mathbf{r}_N \psi^2 \Big/ \int d\mathbf{r}_1 \cdots d\mathbf{r}_N \psi^2. \tag{6}$$

Hereafter $g_2(r_{12})$, also known as the "radial distribution function," will be denoted simply by $g(r)$.

It will be immediately recognized that the relation between $f(r)$ and $g(r)$ given by Eq. (5) is formally identical to the expression for the radial distribution function of a fictitious classical gas interacting via a central pair potential given by $-2\beta^{-1} \ln f(r)$. Consequently, all the methods used in classical statistical mechanics for obtaining the radial distribution function can, in principle, be used in this case. In practice, however, the system is too dense for cluster expansions or approximate integral equations to be particularly useful (Murphy and Watts, 1970; Watts and McGee, 1976). Furthermore, there is great merit in using an exact method to evaluate $\langle H \rangle$, since if one does so one then knows by the variational theorem that the calculated $\langle H \rangle$ should not lie below the experimental value. For this reason, the Monte Carlo method, excellent descriptions of which are given in the review article by Barker and Henderson (1976) and in the book by Watts and McGee (1976), was used in all the calculations presented here.

As was indicated $\langle U_3 \rangle$ is a very small term. Experimentally, the ground-state energy per particle of liquid ^{4}He is (when divided by the Boltzmann constant) about -7.14 K, and if u_3 is taken to be the triple–dipole term, it is found (Murphy and Barker, 1971) that

$$\langle U_3 \rangle / Nk = 0.14 \text{ K}(\rho/\rho_0)^3, \tag{7}$$

where $\rho_0 \sigma^3 = 0.364$ is the equilibrium density of liquid ^{4}He. The smallness of this quantity strongly suggests that it should be entirely adequate to ignore it in first approximation in the Hamiltonian and consequently in the wave function, and to include it only in the final calculations of ground-state energy and other thermodynamic properties, treating it as a perturbation and using the explicit formula given by Eq. (7).

The question now arises as to what form to choose for $f(r)$, the pair wave function. It is fairly clear that $f(r)$ must satisfy two conditions: (1) it must approach zero quite rapidly as $r \to 0$, due to the repulsion of the interatomic potential which will make it quite improbable that particles will approach one another; (2) it must approach a constant (usually taken to be unity) as $r \to \infty$, because of the fact that the system is a fluid and the particles are confined only by the sides of the volume V (in contrast to, e.g., a solid, where the particles are localized at lattice sites).

One procedure (McMillan, 1965; Schiff and Verlet, 1967; Murphy and Watts, 1970) is based on the observation that the WKB solution to the two-body Schrödinger equation using only the repulsive term of the potential is essentially $\exp[-(\zeta/r)^5]$, with $\zeta = \sigma(4m\varepsilon\sigma^2/25\hbar^2)^{1/10} \cong 0.987\sigma$. This suggests taking for $f(r)$

$$f(r) = \exp[-(\alpha/r)^5], \tag{8}$$

with α a variational parameter chosen by computing $\langle H \rangle$ for a number of values of α and taking the value that minimizes the ground-state expectation value of the Hamiltonian. The value of α that minimizes $\langle H \rangle$ turns out to be $\alpha \simeq 1.02\sigma$. The form of $f(r)$ given by Eq. (8) has an additional advantage in that since $f(r) = f(r/\alpha)$, it follows that the calculated $g(r)$ has the following scaling property:

$$g(\rho, r, \alpha) = g(\rho\alpha^3, r/\alpha). \tag{9}$$

This means that once results are computed at one density they can be reused in the calculation at different densities. The calculation thus described leads to fairly good agreement with the experimental ground-state energy and between computed and observed $g(r)$.

A similar, but more general, form of $f(r)$ that preserves the scaling property [Eq. (9)] is the form (Murphy, 1972)

$$f(r) = \exp\left\{-\frac{1}{2}\left[\left(\frac{\alpha}{r}\right)^m + \eta\left(\frac{\alpha}{r}\right)^n\right]\right\}, \tag{10}$$

which has four variational parameters (α, m, n, and η) that can be varied subject to the restriction that m and n be integers with $m > n > 3$, and $\eta = \pm 1$. It was found that the values of the parameters that minimized the expectation value of the Hamiltonian for the Lennard-Jones potential were $\alpha/\sigma = 1.05$; $m = 5$, $n = 4$, and $\eta = +1$. It turned out that the calculated ground-state energy and radial distribution functions were in better agreement with experiment than were calculations using the form for $f(r)$ given in Eq. (8).

Since it is plain that the forms given by Eqs. (8) and (9) are only the WKB solutions and only for the 6-12 potential, an obvious next step (McGee and Murphy, 1972) was to improve the calculation by taking not just the WKB solution but rather the exact solution to the Schrödinger equation and to do so for other forms of the potential. The argument here is quite simple: if one is to assume that the solution to the N-particle Schrödinger equations is to be constructed as a product of factors of pair functions, it seems logical *a priori* that these pair functions ought to be solutions to the two-body Schrödinger equation. This assumption, although it leads to some difficulties with the boundary conditions we must circumvent, is the procedure to be described in this section.

To begin, one writes the Schrödinger equation for the relative motion of two ^{4}He particles of mass m (and hence reduced mass $m/2$) interacting via a central potential $u_2(r)$:

$$(-\hbar^2/m)\nabla^2\psi + u_2(r)\psi = E\psi, \tag{11}$$

where E is the energy eigenvalue. The usual assumption that ψ can be written in the form $R(r)Y_{lm}(\theta, \phi)$, where the Y_{lm} are spherical harmonics, leads to the following ordinary differential equation for R:

$$-\frac{\hbar^2}{m}\left[\frac{d^2R}{dr^2} + \frac{2}{r}\frac{dR}{dr}\right] + \left[u_2(r) + \frac{\hbar^2 l(l+1)}{2mr^2}\right]R = ER.$$

The introduction of the reduced wave function

$$\chi(r) \equiv rR(r)$$

yields the equation

$$(-\hbar^2/m)\chi'' + \left[u_2(r) + \frac{\hbar^2 l(l+1)}{2mr^2}\right]\chi = E\chi.$$

Here l is the quantum number of the square of the angular momentum. The fact that ^{4}He atoms are spinless bosons leads us to consider the s state ($l = 0$), and if we further take $E = 0$, we are led to the following differential equation, which describes the zero-energy, s-state reduced radial wave equation:

$$\chi'' - (m/\hbar^2)u_2(r)\chi(r) = 0. \tag{12}$$

For large values of r, $u_2(r)$ approaches zero rapidly, and so [from Eq. (12)] χ is obviously asymptotically a linear function of r, which is conventionally written in the form

$$\chi(r) = 1 - (r/a) \tag{13}$$

where a, the scattering length, is the negative of the zero-energy s-wave scattering amplitude. Now, near the threshold for a bound state, the behavior of a (Messiah, 1961) is that it is negative if there is no two-body bound state and positive if there is a two-body bound state. But if the depth of the potential $u_2(r)$ is just sufficient to form a bound state with zero binding energy, then a is infinite, and in this case from Eq. (13), $\chi(r) \to 1$ as $r \to \infty$. If there is a two-body bound state with very small binding energy or if the two-body system is unbound but is very close to being bound, then a is finite but very large in absolute value; but unless a is infinite, $|\chi(r)| \to \infty$ as $r \to \infty$.

In the case of ^{4}He the scattering length for any reasonable form of the potential may be expected to be very large in absolute value because the ^{4}He dimer is either just bound or just unbound. Lim *et al.* (1977) have noted some interesting consequences of this fact. For the Lennard-Jones potential, Cashion (1968) has shown that the ^{4}He dimer is very nearly bound; Bruch and McGee (1967, 1970) have shown that for their more realistic forms the ^{4}He dimer is just bound, with binding energies in the range 10^{-4} to 10^{-2} K. But since for all of these potentials the scattering length is finite, not infinite, it is clear that it is necessary to modify χ somehow if it is to be an acceptable Jastrow wave function. The modification that is used is different in the two cases $a > 0$ and $a < 0$.

In the case $a > 0$, where there is a bound state, then $\chi(r)$ has a maximum at a value of r that will be denoted† by r_0. In fact, since a is large, this value of

† In Fig. 1 (and its caption) of the paper by McGee and Murphy (1972), one should substitute r_a for r_0, since r_0 is used elsewhere in that paper in the same context as it is used here.

r_0 is also large, and so one can, simply by normalizing χ such that $\chi(r_0) = 1$, take

$$f(r) = \begin{cases} \chi(r), & 0 < r < r_0 \\ 1, & r_0 \leq r. \end{cases} \tag{14}$$

Thus, $f(r)$ and its first derivative are continuous. Furthermore, $f(r)$ approaches unity asymptotically and is the exact zero-energy two-body s-wave wave function for $0 < r < r_0$.

In the case where there is no bound state ($a < 0$), the procedure differs slightly. Here one constructs a new potential, $u_2^*(r)$, which just has a bound state, (i.e., that potential for which $a \to \infty$) and takes $f(r) = \chi^*(r)$, where χ^* is the solution of the two-body Schrödinger equation (12) with u_2 replaced by u_2^*:

$$(-\hbar^2/m)\chi^{*\prime\prime} + u_2^*(r)\chi^* = 0. \tag{15}$$

The $f(r)$ thus obtained is continuous, has a continuous first derivative, approaches unity asymptotically, and is for small r very close to the solution to the two-body reduced Schrödinger equation, since u_2^* is very close to u_2. In the case of the Lennard-Jones potential, the most straightforward modification of u_2 is simply to multiply the parameter ε by 1.014887, i.e., to increase the depth by about 1.5% while holding the parameter σ constant. The solution to the appropriate differential equation was obtained by standard Runge–Kutta methods, and for the Lennard-Jones case, $f(r)$ is tabulated in Table I.

Given $f(r)$ by the above procedure, $g(r)$, the radial distribution function was calculated by the Monte Carlo method, considering a cubical box of side L with the nearest-image periodic boundary conditions. The box contains $N_s = 32$ particles in volume L^3 with ρ equal to the density of the liquid helium system being simulated. A number of configurations were then generated by the usual biased random walk method with probability proportional to ψ^2, and the radial distribution function computed as $4\pi N_s \rho g(r) r^2 \Delta r = \langle N(r) \rangle$, where $\langle N(r) \rangle$ is the average over the configurations of the number of interparticle distances lying between $r - \Delta r/2$ and $r + \Delta r/2$, with $\Delta r = 0.01\sigma$. The first 16,000 ("equilibrating") configurations were discarded; in most cases the averages were taken over the next 320,000 configurations, but 464,000 were used in some cases. The results for $g(r)$ for the Lennard-Jones potential are contained in Table I for several densities.

It turns out that the $f(r)$ computed for the Lennard-Jones potential from Eq. (15) and tabulated in Table I is very close numerically over the range $0 \leq r/\sigma \leq 2.2$ to the variational form given by Eq. (10) with the parameters

Table I

WAVE FUNCTION AND RADIAL DISTRIBUTION FUNCTION FOR ZERO-TEMPERATURE CASE AS A FUNCTION OF DENSITY

r/σ	$f(r/\sigma)$	$g(r/\sigma)$	$g(r/\sigma)$	$g(r/\sigma)$
		$\rho\sigma^3 = 0.32$	$\rho\sigma^3 = 0.364$	$\rho\sigma^3 = 0.40$
0.675	5.796(10^{-4})	0.0	0.0	0.0
0.685	9.508(16^{-4})	1.65(10^{-4})	0.0	0.0
0.695	1.498(10^{-3})	0.0	0.0	1.29(10^{-4})
0.705	2.276(10^{-3})	9.38(10^{-4})	0.0	7.51(10^{-4})
0.715	3.344(10^{-3})	6.08(10^{-4})	0.0	1.22(10^{-4})
0.725	4.770(10^{-3})	1.47(10^{-4})	3.59(10^{-4})	5.97(10^{-4})
0.735	6.620(10^{-3})	1.44(10^{-4})	4.56(10^{-4})	5.75(10^{-4})
0.745	8.963(10^{-3})	2.80(10^{-4})	1.02(10^{-3})	1.79(10^{-3})
0.755	1.186(10^{-2})	2.04(10^{-3})	1.90(10^{-3})	1.42(10^{-3})
0.765	1.538(10^{-2})	3.32(10^{-3})	4.19(10^{-3})	4.57(10^{-3})
0.775	1.958(10^{-2})	3.88(10^{-3})	5.42(10^{-3})	4.76(10^{-3})
0.785	2.449(10^{-2})	5.17(10^{-3})	6.89(10^{-3})	6.56(10^{-3})
0.795	3.015(10^{-2})	8.12(10^{-3})	8.95(10^{-3})	1.06(10^{-2})
0.805	3.658(10^{-2})	7.91(10^{-3})	1.68(10^{-2})	1.34(10^{-2})
0.815	4.381(10^{-2})	1.32(10^{-2})	2.18(10^{-2})	2.45(10^{-2})
0.825	5.183(10^{-2})	2.25(10^{-2})	2.99(10^{-2})	3.38(10^{-2})
0.835	6.063(10^{-2})	2.65(10^{-2})	3.60(10^{-2})	4.25(10^{-2})
0.845	7.019(10^{-2})	3.67(10^{-2})	5.14(10^{-2})	5.69(10^{-2})
0.855	8.050(10^{-2})	4.83(10^{-2})	6.03(10^{-2})	6.97(10^{-2})
0.865	9.151(10^{-2})	6.07(10^{-2})	7.83(10^{-2})	8.77(10^{-2})
0.875	0.1032	6.85(10^{-2})	9.53(10^{-2})	0.107
0.885	0.1155	8.19(10^{-2})	0.117	0.138
0.895	0.1283	0.115	0.135	0.164
0.905	0.1417	0.129	0.176	0.189
0.915	0.1556	0.159	0.197	0.211
0.925	0.1698	0.175	0.234	0.247
0.935	0.1844	0.205	0.254	0.291
0.945	0.1993	0.240	0.300	0.319
0.955	0.2145	0.266	0.327	0.366
0.965	0.2298	0.297	0.362	0.404
0.975	0.2453	0.334	0.404	0.458
0.985	0.2609	0.365	0.434	0.510
0.995	0.2766	0.402	0.489	0.542
1.005	0.2922	0.445	0.524	0.590
1.015	0.3078	0.466	0.559	0.621
1.025	0.3235	0.509	0.610	0.662
1.035	0.3390	0.558	0.649	0.705
1.045	0.3543	0.578	0.678	0.783
1.055	0.3696	0.635	0.710	0.800
1.065	0.3846	0.671	0.751	0.846
1.075	0.3995	0.690	0.791	0.869
1.085	0.4142	0.708	0.832	0.896
1.095	0.4287	0.769	0.866	0.951
1.105	0.4429	0.780	0.903	0.967
1.115	0.4569	0.819	0.940	1.020
1.125	0.4706	0.847	0.956	1.022
1.135	0.4841	0.893	1.006	1.077
1.145	0.4973	0.916	1.000	1.116
1.155	0.5102	0.929	1.064	1.108
1.165	0.5228	0.959	1.086	1.145
1.175	0.5352	1.014	1.093	1.176
1.185	0.5473	1.017	1.103	1.189
1.195	0.5591	1.017	1.147	1.201

Table I (*continued*)

r/σ	f(r/σ)	g(r/σ) $\rho\sigma^3 = 0.32$	g(r/σ) $\rho\sigma^3 = 0.364$	g(r/σ) $\rho\sigma^3 = 0.40$
1.205	0.5706	1.018	1.148	1.245
1.215	0.5818	1.057	1.171	1.242
1.225	0.5928	1.099	1.179	1.257
1.235	0.6035	1.082	1.195	1.265
1.245	0.6139	1.135	1.216	1.252
1.255	0.6240	1.153	1.215	1.275
1.265	0.6339	1.164	1.224	1.275
1.275	0.6435	1.165	1.236	1.257
1.285	0.6529	1.154	1.237	1.274
1.295	0.6620	1.189	1.234	1.298
1.305	0.6708	1.185	1.251	1.275
1.315	0.6795	1.170	1.265	1.273
1.325	0.6878	1.120	1.259	1.266
1.335	0.6960	1.219	1.253	1.290
1.345	0.7039	1.216	1.240	1.277
1.355	0.7116	1.196	1.245	1.261
1.365	0.7191	1.206	1.248	1.275
1.375	0.7264	1.223	1.246	1.273
1.385	0.7335	1.225	1.255	1.252
1.395	0.7404	1.200	1.238	1.246
1.405	0.7471	1.205	1.233	1.227
1.415	0.7536	1.227	1.245	1.235
1.425	0.7599	1.203	1.215	1.221
1.435	0.7660	1.203	1.216	1.221
1.445	0.7720	1.216	1.232	1.239
1.455	0.7778	1.205	1.209	1.206
1.465	0.7834	1.206	1.200	1.205
1.475	0.7889	1.199	1.210	1.195
1.485	0.7942	1.199	1.182	1.120
1.495	0.7994	1.181	1.180	1.176
1.505	0.8044	1.194	1.172	1.159
1.515	0.8093	1.190	1.160	1.155
1.525	0.8140	1.176	1.173	1.153
1.535	0.8187	1.158	1.152	1.151
1.545	0.8231	1.169	1.144	1.130
1.555	0.8275	1.145	1.140	1.137
1.565	0.8317	1.154	1.137	1.113
1.575	0.8359	1.139	1.130	1.112
1.585	0.8399	1.132	1.117	1.089
1.595	0.8438	1.137	1.122	1.094
1.605	0.8476	1.127	1.105	1.097
1.615	0.8512	1.122	1.096	1.072
1.625	0.8548	1.112	1.090	1.073
1.635	0.8583	1.107	1.076	1.083
1.645	0.8617	1.115	1.067	1.071
1.655	0.8650	1.101	1.070	1.046
1.665	0.8682	1.090	1.075	1.035
1.675	0.8713	1.095	1.044	1.041
1.685	0.8744	1.075	1.040	1.025
1.695	0.8774	1.087	1.034	1.013
1.705	0.8802	1.062	1.029	1.016
1.715	0.8830	1.055	1.029	1.008
1.725	0.8858	1.059	1.018	1.019
1.735	0.8884	1.053	1.031	1.001
1.745	0.8910	1.047	1.018	0.980

(*continued*)

Table I (*continued*)

r/σ	f(r/σ)	g(r/σ)	g(r/σ)	g(r/σ)
		$\rho\sigma^3 = 0.32$	$\rho\sigma^3 = 0.364$	$\rho\sigma^3 = 0.40$
1.755	0.8935	1.045	1.005	0.990
1.765	0.8960	1.043	1.019	0.994
1.775	0.8989	1.024	1.005	0.979
1.785	0.9010	1.025	1.001	0.974
1.795	0.9031	1.019	0.980	0.982
1.805	0.9053	1.025	0.986	0.963
1.815	0.9074	1.027	0.988	0.962
1.825	0.9095	1.004	0.985	0.953
1.835	0.9116	1.002	0.975	0.957
1.845	0.9135	0.995	0.970	0.951
1.855	0.9155	1.004	0.972	0.951
1.865	0.9174	0.999	0.969	0.954
1.875	0.9192	1.000	0.982	0.955
1.885	0.9210	0.988	0.979	0.950
1.895	0.9228	0.971	0.960	0.951
1.905	0.9245	0.979	0.962	0.938
1.915	0.9261	0.982	0.949	0.940
1.925	0.9278	0.986	0.961	0.929
1.935	0.9293	0.984	0.951	0.949
1.945	0.9309	0.975	0.954	0.931
1.955	0.9324	0.975	0.951	0.932
1.965	0.9339	0.958	0.949	0.931
1.975	0.9353	0.971	0.956	0.942
1.985	0.9368	0.961	0.956	0.942
1.995	0.9381	0.967	0.961	0.940
2.005	0.9394	0.957	0.949	0.928
2.015	0.9408	0.957	0.947	0.933
2.025	0.9420	0.959	0.955	0.948
2.035	0.9433	0.962	0.943	0.945
2.045	0.9446	0.961	0.953	0.937
2.055	0.9457	0.954	0.943	0.951
2.065	0.9469	0.968	0.954	0.946
2.075	0.9480	0.949	0.957	0.945
2.085	0.9491	0.968	0.962	0.945
2.095	0.9502	0.958	0.946	0.960
2.105	0.9512	0.958	0.952	0.948
2.115	0.9523	0.960	0.955	0.957
2.125	0.9533	0.956	0.960	0.951
2.135	0.9543	0.969	0.952	0.959
2.145	0.9553	0.962	0.955	0.955
2.155	0.9562	0.958	0.960	
2.165	0.9571	0.965	0.969	
2.175	0.9580	0.965	0.965	
2.185	0.9589	0.965	0.966	
2.195	0.9597	0.966	0.962	
2.205	0.9606	0.963	0.960	
2.215	0.9614	0.976		
2.225	0.9622	0.963		
2.235	0.9630	0.972		
2.245	0.9638	0.963		
2.255	0.9645	0.957		
2.265	0.9653	0.970		
2.275	0.9660	0.971		
2.285	0.9667	0.967		
2.295	0.9674	0.973		
2.305	0.9681	0.975		

listed that minimize $\langle H \rangle$ for the Lennard-Jones potential. It is not surprising, therefore, that the computed $g(r)$ from the two cases are also practically identical. The significance of this remark goes beyond merely the observation that one has in Eq. (10) a reasonably good analytic representation of $\chi^*(r)$, the solution to Eq. (15). It means that the zero-parameter procedure described here, namely, use of the zero-energy s-wave solutions to the Schrödinger equation as a Jastrow function, did as well in calculating the radial distribution function as did the four-variational parameter form represented by Eq. (10). Except for some calculations using eight parameters (DeMichelis and Reatto, 1974), the results we have presented represent the best agreement with the experimental measurements of $g(r)$ taken by Achter and Meyer (1969), and the $g(r)$ computed for different forms of $u_2(r)$ (Murphy and McGee, 1973) are in better agreement yet with experiment (although not quite as good as the best results of DeMichelis and Reatto). In addition, by the variational theorem, one has an upper bound to the ground-state energy, i.e., this procedure is in effect a variational calculation without any variational parameters. Results for $g(r)$ for forms of the potential other than the Lennard-Jones and of thermodynamic properties will be published elsewhere (McGee and Murphy, 1977).

IV. Finite-Temperature Approach: The Effective Potential Approximation

The usual method of formulating quantum statistical mechanics is by means of the density matrix. In the canonical ensemble, one begins with the Bloch equation, which is formally identical to the time-dependent Schrödinger equation for imaginary times:

$$\hat{H}\hat{\rho} + (\partial\hat{\rho}/\partial\beta) = 0 \tag{16}$$

with the boundary condition that $\hat{\rho}(0) = 1$. A normalized formal solution to Eq. (16) is

$$\hat{\rho} = e^{-\beta\hat{H}}/Z, \tag{17}$$

where $Z = \mathrm{Tr}(e^{-\beta\hat{H}})$. The factor $Z(T, V, N)$, which serves to normalize $\hat{\rho}$, is in fact the canonical partition function, which is related to the Helmholtz free energy $F(T, V, N)$ in the usual fashion:

$$F(T, V, N) = -kT \ln Z(T, V, N).$$

Given the Helmholtz free energy, all the other thermodynamic properties can in principle be calculated from macroscopic thermodynamic relations, just as is done in classical statistical mechanics.

The density operator can be represented by a square Hermitian matrix.

For an arbitrary complete orthonormal set of basis functions $|n\rangle$, the elements of the density matrix are simply

$$\rho_{mn} = \langle m|\hat{\rho}|n\rangle.$$

Specifically, if the set $|n\rangle$ are chosen to be energy eigenfunctions, $\hat{H}|n\rangle = E_n|n\rangle$, then the density matrix is diagonal, i.e.,

$$\rho_{mn} = Z^{-1}e^{-\beta E_m}\,\delta_{mn},$$

and

$$Z = \sum_m e^{-\beta E_m}.$$

A more useful representation of the density matrix is the "coordinate representation," in which for a system of N particles

$$\rho_{\mathbf{rr}} \equiv \rho(\mathbf{r}_1 \cdots \mathbf{r}_N, \mathbf{r}_1' \cdots \mathbf{r}_N') = Z^{-1} \sum_m \phi_m^*(\mathbf{r}')e^{-\beta\hat{H}}\phi_m(\mathbf{r}). \tag{18}$$

Here the set of functions ϕ_m are any complete, orthonormal, properly symmetrized set of functions. In this representation, the diagonal element $\rho_{\mathbf{rr}}$, which in the density matrix formulation represents the probability density for the configuration $\mathbf{r}_1, \mathbf{r}_2, \ldots \mathbf{r}_N$, is given by

$$\rho_{\mathbf{rr}} = Z^{-1} \sum_n \phi_n^*(\mathbf{r}_1, \mathbf{r}_2, \ldots \mathbf{r}_N)e^{-\beta\hat{H}}\phi_n(\mathbf{r}_1, \mathbf{r}_2, \ldots \mathbf{r}_N). \tag{19}$$

The Slater sum $W(\mathbf{r}_1, \mathbf{r}_2, \ldots \mathbf{r}_N)$ is now defined by

$$W(\mathbf{r}_1, \mathbf{r}_2, \ldots \mathbf{r}_N) = N!\lambda^{3N} \sum_n \phi_n^*(\mathbf{r}_1, \mathbf{r}_2, \ldots \mathbf{r}_N)e^{-\beta\hat{H}}\phi_n(\mathbf{r}_1, \mathbf{r}_2, \ldots \mathbf{r}_N) \tag{20}$$

where $\lambda = h(2\pi mkT)^{-1/2}$ is the usual thermal deBroglie wavelength of a particle. From Eqs. (17) and (18), it follows that

$$\rho_{\mathbf{rr}} = \frac{1}{\lambda^{3N}N!Z} W(\mathbf{r}_1, \mathbf{r}_2, \ldots \mathbf{r}_N). \tag{21}$$

It is immediately obvious from Eq. (21) that the Slater sum is the analog of the Boltzmann factor $\exp[-\sum u_2(r_{ij})]$, since if one replaces the Slater sum in Eq. (21) by the Boltzmann factor one has one of the basic equations of classical statistical mechanics. In analogy with classical statistical mechanics, it follows that one can define the radial distribution function (and in principle three-body, etc., distribution functions also, although these are not of interest at present) by

$$g(r_{12}) = g(|\mathbf{r}_1 - \mathbf{r}_2|) = N(N-1)\rho^{-2} \int d\mathbf{r}_3 \cdots d\mathbf{r}_N\, W \bigg/ \int d\mathbf{r}_1 \cdots d\mathbf{r}_N\, W. \tag{22}$$

In fact, the analogy represented by Eq. (21) between the Slater sum and the Boltzmann factor is not merely a formal one. In the classical region (i.e., in the region where $\rho\lambda^3$ is small), it is known [see, for example, Hirschfelder *et al.* (1954)] that the Slater sum approaches the Boltzmann factor as $\rho\lambda^3 \to 0$. Consequently, one has in principle a prescription for computing the thermodynamic properties of a quantum mechanical system and the guarantee that the procedure goes over to classical statistical mechanics in the classical limit $\hbar \to 0$ or $T \to \infty$. Deviations from classical behavior come from two sources: those that are usually termed "diffraction effects," which are a consequence of the wave nature of the particles, and those termed "exchange effects," which are a consequence of the requirement that the wave function be symmetric (for bosons) or antisymmetric (for fermions) under exchange of particles.

The above formulation is formally exact, but as it stands it would be quite difficult to apply to real systems. The reason for this is that while it is a textbook problem (see, for example, Münster, 1972) to display a complete set of properly symmetrized eigenfunctions ϕ_m as called for by Eq. (20) if the particles are noninteracting, it is impossible to display such a set of ϕ_m (or even one of them, for that matter) if the system is interacting. Accordingly, one is obliged to use some sort of approximation procedure, and one such procedure, the effective potential approximation, will be described here.

Precisely the same sort of argument that led to the Jastrow wave function [Eq. (3)] leads us to the following factorization of the Slater sum:

$$W(\mathbf{r}_1, \mathbf{r}_2, \ldots \mathbf{r}_N) = \prod_{i<j} W_2(r_{ij}) \prod_{i<j<k} W_3(\mathbf{r}_i, \mathbf{r}_j, \mathbf{r}_k) \cdots . \tag{23}$$

Here the quantity $W_2(r)$ is the Slater sum for two particles, W_3 is that portion of the three-particle Slater sum that cannot be accounted for in terms of two-particle Slater sums, etc. Truncation of the expansion represented by Eq. (23) at the first term, i.e., writing

$$W(\mathbf{r}_1, \mathbf{r}_2, \ldots \mathbf{r}_N) = \prod_{i<j} W_2(r_{ij}) = \exp\left[\sum_{i<j} w(r_{ij})\right] \tag{24}$$

is called the effective potential approximation, since in this approximation the radial distribution function $g(r)$ can be seen from Eq. (22) to be equal to the radial distribution function of a fictitious classical gas interacting via a pair potential $-kTw(r)$. As noted in Section III, the formal equivalence of the expression for $g(r)$ to the expressions found in classical statistical mechanics means that all the procedures used there to calculate $g(r)$ from a given $w(r)$ can be used. At high temperatures, i.e., above the critical point, ^{4}He is in a gaseous phase with relatively low density, and for this reason approximate integral equation schemes may be expected to be more useful here than in the zero-temperature liquid problem. Two of the most widely used approximate integral equations are the Percus–Yevick equation (Percus and Yevick, 1958) and the hypernetted chain equation (van Leeuwen *et al.*, 1959), which are discussed in some detail in a review article by Watts (1973). The Percus–Yevick and hypernetted chain equations were used fairly successfully to calculate $g(r)$ and hence the critical point parameters for helium by Bruch and McGee (1975) and by Bruch *et al.* (1974).

In calculating the Slater sums $W_2(r)$, the method and program of Klemm and Storer (1973) was used. At the temperatures involved, namely 7 K and 8 K, the exchange terms are small and were neglected. The procedure of Klemm and Storer consists of an expansion of the quantity $W_2(r)$ into a sum of contributions from all partial waves, with the Hamiltonian operator for the lth partial wave being expressed as a sum of two noncommuting operators, each of which obeys an equation like the Bloch equation (16). The density operator is then treated by a high-temperature expansion, and the resulting high temperature approximation is used as a starting point for an iterative procedure to find the density matrix at lower temperatures.

The Slater sums thus computed were obtained for both temperatures at spacings of 0.1σ (although a finer spacing was used in the initial high-temperature iterations). In order to interpolate in the table thus obtained, a different procedure was used in each of three different regions, as follows. In the region $0 < r/\sigma < 0.7$, the Slater sum was fitted by the expression

$$w(r) = A - (\gamma\sigma/r)^5 - (\gamma\sigma/r)^4. \tag{25}$$

In the region $0.7 < r/\sigma < 2.2$, purely numerical (Aitkin) interpolation methods were used. For $r/\sigma > 2.05$, the semiclassical expansion is valid:

$$\begin{aligned} \ln w(r) &\cong -\beta u_2(r) - (\beta^2 h^2/6m)\nabla^2 u_2(r) \\ &= -C[(\sigma/r)^{12} - (\sigma/r)^6] - D[22(\sigma/r)^{14} - r(\sigma/r)^8]. \end{aligned} \tag{26}$$

The values of the parameters A, γ, C, and D are contained in Table II, for the two temperatures 7 K and 8 K. The resultant values of $w(r)$ are tabulated in Table III for $T = 7$ K and in Table IV for $T = 8$ K.

Table II

PARAMETERS FOR CURVE FITS OF SLATER SUMS[a]

Temperature	7 K	8 K
A	2.469	2.469
γ	1.061	1.061
C	5.8400	5.1100
D	1.5476	1.1849

[a] See Eqs. (25) and (26).

Table III

Slater Sum and Radial Distribution Function for $T = 7$ K

r/σ	$W_2(r/\sigma)$	$g(r/\sigma)$					
		$\rho\sigma^3 = 0.175$	$\rho\sigma^3 = 0.200$	$\rho\sigma^3 = 0.225$	$\rho\sigma^3 = 0.25$	$\rho\sigma^3 = 0.30$	$\rho\sigma^3 = 0.35$
0.705	$2.18(10^{-5})$	0.0	0.0	0.0	0.0	0.0	0.0
0.715	$5.01(10^{-5})$	$1.39(10^{-3})$	0.0	$2.76(10^{-4})$	$9.72(10^{-4})$	0.0	0.0
0.725	$1.08(10^{-4})$	0.0	$1.19(10^{-3})$	0.0	$7.09(10^{-4})$	0.0	$1.69(10^{-4})$
0.735	$2.20(10^{-4})$	$3.29(10^{-4})$	$8.63(10^{-4})$	0.0	$2.30(10^{-4})$	0.0	$1.64(10^{-3})$
0.745	$4.26(10^{-4})$	0.0	$2.80(10^{-4})$	$4.98(10^{-4})$	$2.24(10^{-4})$	$1.31(10^{-3})$	$9.60(10^{-4})$
0.755	$7.83(10^{-4})$	$9.35(10^{-4})$	$2.18(10^{-3})$	$2.18(10^{-3})$	$1.75(10^{-3})$	$1.82(10^{-4})$	$2.03(10^{-3})$
0.765	$1.38(10^{-3})$	$1.52(10^{-3})$	$4.51(10^{-2})$	$4.72(10^{-3})$	$2.55(10^{-3})$	$5.13(10^{-3})$	$3.04(10^{-3})$
0.775	$2.33(10^{-3})$	$4.44(10^{-3})$	$2.33(10^{-3})$	$4.37(10^{-3})$	$2.28(10^{-3})$	$5.00(10^{-3})$	$6.36(10^{-3})$
0.785	$3.80(10^{-3})$	$3.17(10^{-3})$	$5.55(10^{-3})$	$2.47(10^{-3})$	$5.85(10^{-3})$	$5.72(10^{-3})$	$8.94(10^{-3})$
0.795	$5.99(10^{-3})$	$4.50(10^{-3})$	$1.23(10^{-2})$	$7.43(10^{-3})$	$1.32(10^{-2})$	$1.39(10^{-2})$	$1.42(10^{-2})$
0.805	$9.13(10^{-3})$	$1.26(10^{-2})$	$1.42(10^{-2})$	$1.07(10^{-2})$	$1.32(10^{-2})$	$1.01(10^{-2})$	$2.29(10^{-2})$
0.815	$1.31(10^{-2})$	$1.95(10^{-2})$	$1.83(10^{-2})$	$1.56(10^{-2})$	$2.32(10^{-2})$	$1.92(10^{-2})$	$3.06(10^{-2})$
0.825	$1.76(10^{-2})$	$2.64(10^{-2})$	$3.43(10^{-2})$	$2.74(10^{-2})$	$3.16(10^{-2})$	$2.89(10^{-2})$	$3.37(10^{-2})$
0.835	$2.34(10^{-2})$	$3.92(10^{-2})$	$2.72(10^{-2})$	$3.41(10^{-2})$	$3.30(10^{-2})$	$3.37(10^{-2})$	$5.03(10^{-2})$
0.845	$3.07(10^{-2})$	$4.73(10^{-2})$	$3.79(10^{-2})$	$5.15(10^{-2})$	$4.44(10^{-2})$	$5.40(10^{-2})$	$6.06(10^{-2})$
0.855	$3.94(10^{-2})$	$5.66(10^{-2})$	$6.53(10^{-2})$	$5.35(10^{-2})$	$6.21(10^{-2})$	$7.36(10^{-2})$	$8.00(10^{-2})$
0.865	$4.99(10^{-2})$	$5.77(10^{-2})$	$6.96(10^{-2})$	$7.27(10^{-2})$	$6.90(10^{-2})$	$9.03(10^{-2})$	$8.94(10^{-2})$
0.875	$6.27(10^{-2})$	$9.26(10^{-2})$	$7.73(10^{-2})$	$8.66(10^{-2})$	$8.80(10^{-2})$	0.110	0.114
0.885	$7.84(10^{-2})$	0.117	0.128	0.119	0.114	0.127	0.138
0.895	$9.79(10^{-2})$	0.116	0.120	0.144	0.154	0.153	0.184
0.905	0.1224	0.167	0.160	0.192	0.175	0.203	0.207
0.915	0.1452	0.163	0.191	0.205	0.212	0.222	0.248
0.925	0.1633	0.197	0.218	0.233	0.223	0.245	0.272
0.935	0.1854	0.228	0.234	0.249	0.269	0.277	0.315
0.945	0.2115	0.242	0.265	0.281	0.286	0.334	0.342
0.955	0.2387	0.297	0.321	0.333	0.345	0.338	0.364
0.965	0.2688	0.313	0.349	0.353	0.365	0.396	0.448
0.975	0.3026	0.360	0.344	0.403	0.394	0.437	0.456
0.985	0.3413	0.414	0.411	0.434	0.442	0.473	0.550
0.995	0.3865	0.438	0.483	0.469	0.510	0.524	0.576
1.005	0.4392	0.535	0.533	0.529	0.533	0.635	0.639
1.015	0.4831	0.531	0.594	0.594	0.627	0.671	0.694

(continued)

Table III (*continued*)

r/σ	$W_2(r/\sigma)$	$g(r/\sigma)$					
		$\rho\sigma^3 = 0.175$	$\rho\sigma^3 = 0.200$	$\rho\sigma^3 = 0.225$	$\rho\sigma^3 = 0.25$	$\rho\sigma^3 = 0.30$	$\rho\sigma^3 = 0.35$
1.025	0.5134	0.575	0.612	0.657	0.657	0.668	0.757
1.035	0.5397	0.645	0.675	0.644	0.679	0.712	0.757
1.045	0.5908	0.673	0.716	0.728	0.736	0.757	0.784
1.055	0.6302	0.719	0.749	0.730	0.758	0.782	0.838
1.065	0.6705	0.743	0.785	0.763	0.791	0.847	0.884
1.075	0.7120	0.794	0.827	0.820	0.804	0.904	0.920
1.085	0.7551	0.862	0.873	0.873	0.858	0.935	1.018
1.095	0.8005	0.882	0.919	0.894	0.910	0.979	0.991
1.105	0.8480	0.944	0.960	0.928	1.030	0.971	1.051
1.115	0.8888	0.956	0.974	0.988	1.032	1.014	1.097
1.125	0.9212	1.008	1.021	0.998	1.040	1.086	1.126
1.135	0.9555	0.975	1.012	1.056	1.045	1.103	1.129
1.145	0.9911	1.106	1.108	1.062	1.103	1.145	1.178
1.155	1.0247	1.091	1.075	1.085	1.134	1.149	1.172
1.165	1.0575	1.126	1.149	1.165	1.142	1.167	1.215
1.175	1.0899	1.173	1.114	1.154	1.171	1.197	1.260
1.185	1.1219	1.193	1.120	1.171	1.179	1.190	1.254
1.195	1.1539	1.141	1.218	1.190	1.163	1.228	1.274
1.205	1.1859	1.241	1.179	1.243	1.255	1.255	1.262
1.215	1.2130	1.252	1.231	1.268	1.254	1.282	1.304
1.225	1.2349	1.201	1.256	1.286	1.222	1.314	1.301
1.235	1.2570	1.223	1.241	1.236	1.301	1.255	1.315
1.245	1.2787	1.272	1.275	1.268	1.270	1.318	1.331
1.255	1.2987	1.314	1.268	1.275	1.309	1.327	1.326
1.265	1.3175	1.314	1.288	1.312	1.300	1.289	1.335
1.275	1.3354	1.342	1.273	1.299	1.282	1.307	1.342
1.285	1.3524	1.339	1.328	1.309	1.286	1.332	1.335
1.295	1.3687	1.381	1.338	1.347	1.344	1.325	1.331
1.305	1.3844	1.379	1.342	1.365	1.332	1.316	1.364
1.315	1.3970	1.345	1.353	1.362	1.341	1.336	1.335
1.325	1.4065	1.350	1.385	1.338	1.317	1.299	1.342
1.335	1.4156	1.345	1.342	1.305	1.308	1.327	1.344
1.345	1.4241	1.383	1.294	1.362	1.352	1.332	1.316

1.355	1.4313	1.379	1.351	1.361	1.277	1.318	1.294
1.365	1.4376	1.375	1.355	1.348	1.303	1.321	1.332
1.375	1.4431	1.339	1.336	1.341	1.297	1.317	1.322
1.385	1.4479	1.371	1.342	1.344	1.334	1.299	1.296
1.395	1.4521	1.347	1.355	1.331	1.317	1.334	1.299
1.405	1.4557	1.351	1.349	1.308	1.303	1.284	1.281
1.415	1.4575	1.381	1.343	1.332	1.273	1.283	1.292
1.425	1.4576	1.319	1.324	1.309	1.314	1.320	1.240
1.435	1.4574	1.326	1.289	1.302	1.268	1.286	1.280
1.445	1.4567	1.334	1.344	1.308	1.285	1.277	1.238
1.455	1.4553	1.313	1.325	1.292	1.263	1.274	1.253
1.465	1.4534	1.340	1.315	1.284	1.236	1.244	1.236
1.475	1.4509	1.326	1.309	1.305	1.266	1.213	1.228
1.485	1.4481	1.294	1.298	1.263	1.246	1.219	1.191
1.495	1.4449	1.356	1.312	1.254	1.277	1.224	1.191
1.505	1.4414	1.318	1.283	1.251	1.232	1.208	1.182
1.515	1.4370	1.304	1.285	1.309	1.222	1.202	1.190
1.525	1.4317	1.291	1.279	1.255	1.216	1.192	1.177
1.535	1.4263	1.278	1.261	1.228	1.189	1.172	1.160
1.545	1.4207	1.266	1.258	1.212	1.191	1.147	1.131
1.555	1.4148	1.279	1.244	1.214	1.198	1.148	1.131
1.565	1.4086	1.235	1.238	1.221	1.181	1.164	1.124
1.575	1.4023	1.246	1.219	1.217	1.180	1.163	1.124
1.585	1.3958	1.233	1.194	1.222	1.183	1.147	1.108
1.595	1.3892	1.223	1.189	1.182	1.154	1.122	1.107
1.605	1.3826	1.252	1.212	1.178	1.145	1.117	1.100
1.615	1.3756	1.232	1.206	1.206	1.154	1.126	1.102
1.625	1.3682	1.215	1.194	1.165	1.114	1.102	1.062
1.635	1.3608	1.203	1.171	1.141	1.141	1.095	1.072
1.645	1.3534	1.209	1.178	1.148	1.123	1.067	1.065
1.655	1.3460	1.195	1.169	1.126	1.116	1.081	1.070
1.665	1.3385	1.171	1.148	1.128	1.086	1.073	1.054
1.675	1.3311	1.149	1.124	1.131	1.102	1.080	1.042
1.685	1.3238	1.141	1.136	1.143	1.085	1.059	1.019
1.695	1.3165	1.150	1.118	1.099	1.126	1.042	1.032
1.705	1.3092	1.111	1.138	1.113	1.079	1.067	1.006
1.715	1.3019	1.130	1.107	1.088	1.059	1.031	1.005
1.725	1.2944	1.144	1.109	1.082	1.091	1.013	0.986

(continued)

Table III *(continued)*

r/σ	$W_2(r/\sigma)$	$g(r/\sigma)$					
		$\rho\sigma^3 = 0.175$	$\rho\sigma^3 = 0.200$	$\rho\sigma^3 = 0.225$	$\rho\sigma^3 = 0.25$	$\rho\sigma^3 = 0.30$	$\rho\sigma^3 = 0.35$
1.735	1.2871	1.137	1.113	1.089	1.062	1.027	0.984
1.745	1.2799	1.111	1.110	1.092	1.028	1.042	0.979
1.755	1.2727	1.088	1.078	1.092	1.032	1.013	0.976
1.765	1.2657	1.110	1.075	1.079	1.027	1.007	0.987
1.775	1.2587	1.116	1.082	1.061	1.027	1.009	0.983
1.785	1.2519	1.105	1.060	1.069	1.029	0.978	0.973
1.795	1.2453	1.089	1.045	1.041	1.017	0.988	0.948
1.805	1.2388	1.064	1.031	1.037	1.023	0.999	0.974
1.815	1.2323	1.094	1.062	1.029	1.020	0.980	0.968
1.825	1.2258	1.081	1.061	1.026	0.996	0.988	0.949
1.835	1.2195	1.088	1.065	1.025	1.011	0.987	0.943
1.845	1.2133	1.050	1.050	1.050	1.013	0.977	0.942
1.855	1.2073	1.058	1.037	1.019	1.020	0.958	0.945
1.865	1.2014	1.047	1.035	1.016	0.997	0.958	0.937
1.875	1.1956	1.047	1.036	1.022	0.988	0.948	0.953
1.885	1.1900	1.067	1.031	0.991	0.999	0.972	0.943
1.895	1.1846	1.042	1.028	0.998	0.993	0.978	0.955
1.905	1.1793	1.057	1.006	1.001	0.988	0.954	0.941
1.915	1.1741	1.043	1.039	0.983	0.997	0.958	0.948
1.925	1.1689	1.041	1.017	1.011	0.988	0.965	0.941
1.935	1.1639	1.035	1.040	0.994	0.979	0.967	0.925
1.945	1.1590	1.013	1.013	0.999	0.975	0.960	0.934
1.955	1.1543	1.031	1.003	0.985	0.961	0.953	0.939
1.965	1.1497	1.002	1.008	0.982	0.995	0.950	0.942
1.975	1.1452	1.004	0.988	0.985	0.957	0.943	0.923
1.985	1.1409	1.021	1.002	1.001	0.993	0.955	0.930
1.995	1.1367	1.016	1.008	1.002	0.972	0.950	0.933
2.005	1.1327	1.016	0.976	0.990	0.975	0.956	0.935
2.015	1.1287	1.011	0.993	0.984	0.976	0.951	0.931
2.025	1.1248	1.023	1.011	0.979	0.959	0.947	0.934
2.035	1.1210	1.007	0.995	0.975	0.977	0.951	0.931
2.045	1.1173	1.008	0.977	0.988	0.963	0.954	0.943
2.055	1.1138	1.004	0.997	0.976	0.959	0.951	0.950

2.065	1.1104	1.014	0.996	0.986	0.965	0.962	0.931
2.075	1.1070	0.996	0.997	0.983	0.966	0.969	0.938
2.085	1.1038	0.997	1.002	0.965	0.990	0.955	0.950
2.095	1.1007	0.988	0.970	0.980	0.969	0.957	0.944
2.105	1.0977	0.990	0.989	0.988	0.968	0.964	0.962
2.115	1.0948	0.994	0.971	0.992	0.958	0.952	0.965
2.125	1.0919	0.995	0.999	1.002	0.952	0.957	0.970
2.135	1.0892	1.006	0.981	0.976	0.985	0.966	0.956
2.145	1.0865	0.993	1.005	0.993	0.962	0.974	0.936
2.155	1.0839	0.993	0.972	0.980	0.961	0.969	0.950
2.165	1.0814	0.991	0.995	0.980	0.971	0.978	0.967
2.175	1.0790	1.003	0.971	0.985	0.980	0.986	0.966
2.185	1.0767	0.988	0.995	1.004	0.971	0.969	0.974
2.195	1.0745	0.993	0.979	0.996	0.987	0.970	0.966
2.205	1.0723	1.008	0.996	0.985	0.977	0.962	0.955
2.215	1.0702	1.001	0.978	0.997	0.979	0.981	0.974
2.225	1.0681	1.015	0.999	0.971	1.002	0.972	0.974
2.235	1.0661	1.027	1.004	0.997	0.993	0.975	0.971
2.245	1.0642	0.991	0.996	0.987	0.984	0.981	0.977
2.255	1.0623	1.012	0.999	0.991	0.989	0.982	0.980
2.265	1.0605	0.999	1.009	0.999	0.977	0.985	0.986
2.275	1.0588	0.998	1.007	0.991	0.988	0.998	0.993
2.285	1.0571	1.004	1.000	0.980	1.001	0.987	0.992
2.295	1.0555	1.026	1.006	0.986	0.993	0.990	0.986
2.305	1.0539	1.027	1.019	0.989	0.987	0.996	0.981
2.315	1.0524	1.010	0.997	1.007	0.995	0.989	0.979
2.325	1.0509	1.018	1.007	0.998	0.989	0.993	0.993
2.335	1.0495	1.003	0.987	1.007	0.994	0.999	0.998
2.345	1.0481	1.008	1.004	0.998	0.983	0.992	1.001
2.355	1.0468	1.002	1.005	1.012	0.992	0.997	0.998
2.365	1.0455	1.017	1.007	1.002	0.998	0.994	1.000
2.375	1.0442	1.013	1.014	1.004	1.000	1.012	1.003
2.385	1.0430	0.998	0.995	1.008	1.004	0.990	1.008
2.395	1.0418	1.001	1.018	1.010	1.006	1.008	1.012
2.405	1.0407	1.010	1.016	1.012	0.994	1.005	1.016
2.415	1.0396	1.013	1.035	1.021	1.002	1.005	1.019
2.425	1.0385	0.999	1.010	1.021	1.007	1.015	1.011
2.435	1.0375	1.023	1.022	1.016	1.011	1.012	1.013
2.445	1.0365	1.014	1.015	0.988	1.011	0.988	1.017

(continued)

Table III *(continued)*

r/σ	$W_2(r/\sigma)$	$g(r/\sigma)$					
		$\rho\sigma^3 = 0.175$	$\rho\sigma^3 = 0.200$	$\rho\sigma^3 = 0.225$	$\rho\sigma^3 = 0.25$	$\rho\sigma^3 = 0.30$	$\rho\sigma^3 = 0.35$
2.455	1.0355	1.030	1.014	1.005	1.012	0.995	1.009
2.465	1.0346	1.016	1.025	1.011	1.017	1.012	1.018
2.475	1.0337	1.013	1.012	1.003	1.007	1.014	1.007
2.485	1.0328	1.028	1.023	1.019	1.009	1.011	1.008
2.495	1.0319	1.023	1.018	1.022	1.001	1.006	1.016
2.505	1.0311	1.026	1.007	1.011	1.013	1.015	1.023
2.515	1.0303	1.028	1.029	1.018	1.010	1.027	1.018
2.525	1.0295	1.022	1.027	1.003	1.017	1.014	1.019
2.535	1.0287	1.032	1.012	1.013	1.016	1.005	1.019
2.545	1.0280	0.998	1.015	1.013	1.016	1.016	1.013
2.555	1.0273	1.026	1.013	1.019	1.012	1.015	1.018
2.565	1.0266	1.025	1.024	1.023	1.000	0.999	1.028
2.575	1.0259	1.029	1.011	1.026	1.014	1.013	1.015
2.585	1.0253	1.020	1.035	1.009	0.995	1.008	1.018
2.595	1.0246	1.029	1.008	1.024	1.005	1.007	1.022
2.605	1.0240	1.016	1.021	1.017	1.011	1.012	1.027
2.615	1.0234	1.010	1.005	1.018	1.001	1.015	1.011
2.625	1.0228	1.015	1.021	1.012	1.017	1.001	1.016
2.635	1.0223	1.014	1.018	1.017	1.017	1.009	1.014
2.645	1.0217	1.019	1.012	1.009	0.999	1.021	1.011
2.655	1.0212	1.022	1.024	1.027	1.017	1.008	1.016
2.665	1.0207	1.023	1.010	1.022	1.009	1.007	1.020
2.675	1.0202	1.018	1.007	1.018	1.017	1.016	1.020
2.685	1.0197	1.026	1.012	1.019	1.011	1.015	1.022
2.695	1.0192	1.025	1.019	1.025	1.009	1.021	1.014
2.705	1.0188	1.029	1.022	1.024	1.000	1.011	1.019
2.715	1.0183	1.026	1.022	1.017	1.019	1.015	1.016
2.725	1.0179	1.033	1.031	1.010	1.004	1.016	1.008
2.735	1.0175	1.027	1.017	1.019	1.004	0.998	1.030
2.745	1.0171	1.012	1.035	1.011	1.016	1.013	1.009
2.755	1.0167	1.027	1.026	1.020	1.023	1.020	1.018
2.765	1.0163	1.014	1.022	1.028	1.012	1.020	1.020
2.775	1.0159	1.016	1.017	1.019	1.006	1.022	1.014

2.785	1.0155	1.027	1.020	1.010	1.007	1.023	1.001
2.795	1.0152	1.011	1.022	1.012	1.007	1.011	1.002
2.805	1.0148	1.006	1.008	1.015	1.006	0.998	1.005
2.815	1.0145	1.026	1.017	1.014	1.005	1.010	1.007
2.825	1.0142	1.008	1.019	1.007	1.009	1.015	1.013
2.835	1.0138	1.031	1.013	1.022	1.020	1.004	1.012
2.845	1.0135	1.028	1.030	1.009	1.025	1.018	1.009
2.855	1.0132	1.030	1.012	1.017	1.014	0.998	1.006
2.865	1.0129	1.012	1.016	1.007	1.004	0.995	1.006
2.875	1.0126	1.010	1.008	1.014	1.004	1.018	1.003
2.885	1.0123	1.014	1.029	1.012	1.010	1.011	1.001
2.895	1.0121	1.019	1.008	1.010	0.999	1.007	1.004
2.905	1.0118	1.012	1.008	0.996	1.009	1.010	0.996
2.915	1.0115	1.026	1.008	1.019	1.010	1.002	0.996
2.925	1.0113	1.017	1.019	1.026	0.993	0.999	1.006
2.935	1.0110	1.017	1.036	1.017	1.012	1.011	1.001
2.945	1.0108	1.023	1.021	1.011	1.012	1.006	0.998
2.955	1.0106	1.009	1.011	1.008	0.990	0.996	0.994
2.965	1.0103	1.015	1.022	1.005	1.009	0.988	1.004
2.975	1.0101	1.010	1.012	1.015	1.010	0.997	0.997
2.985	1.0099	1.022	1.006	1.008	1.005	0.997	0.994
2.995	1.0097	1.016	1.015	1.002	1.008	1.001	0.990
3.005	1.0095	1.008	1.011	1.006	1.006	0.997	0.991

Table IV

Slater Sum and Radial Distribution for $T = 8$ K

r/σ	$W_2(r/\sigma)$	$g(r/\sigma)$					
		$\rho\sigma^3=0.175$	$\rho\sigma^3=0.20$	$\rho\sigma^3=0.225$	$\rho\sigma^3=0.250$	$\rho\sigma^3=0.30$	$\rho\sigma^3=0.35$
0.705	$2.18(10^{-5})$	$2.57(10^{-4})$	0.0	0.0	$5.00(10^{-4})$	0.0	$3.57(10^{-4})$
0.715	$5.01(10^{-5})$	0.0	$3.04(10^{-4})$	$2.70(10^{-4})$	$4.86(10^{-4})$	0.0	0.0
0.725	$1.08(10^{-4})$	0.0	$8.87(10^{-4})$	0.0	$7.10(10^{-4})$	$5.91(10^{-4})$	0.0
0.735	$2.20(10^{-4})$	$3.29(10^{-4})$	0.0	0.0	$1.84(10^{-3})$	$1.34(10^{-3})$	$6.58(10^{-4})$
0.745	$4.26(10^{-4})$	$2.56(10^{-3})$	$2.24(10^{-3})$	$2.24(10^{-3})$	$1.79(10^{-3})$	$2.05(10^{-3})$	$2.08(10^{-3})$
0.755	$7.83(10^{-4})$	$9.35(10^{-4})$	$1.09(10^{-3})$	$1.94(10^{-3})$	$5.24(10^{-3})$	$4.90(10^{-3})$	$3.43(10^{-3})$
0.765	$1.38(10^{-3})$	$1.52(10^{-3})$	$3.19(10^{-3})$	$4.72(10^{-4})$	$1.49(10^{-3})$	$5.31(10^{-3})$	$3.19(10^{-3})$
0.775	$2.33(10^{-3})$	$4.14(10^{-3})$	$5.43(10^{-3})$	$4.83(10^{-3})$	$6.21(10^{-3})$	$3.62(10^{-3})$	$4.29(10^{-3})$
0.785	$3.80(10^{-3})$	$3.46(10^{-3})$	$1.06(10^{-2})$	$6.95(10^{-3})$	$7.06(10^{-3})$	$9.25(10^{-3})$	$7.93(10^{-3})$
0.795	$5.99(10^{-3})$	$6.75(10^{-3})$	$7.62(10^{-3})$	$9.62(10^{-3})$	$1.50(10^{-2})$	$1.11(10^{-2})$	$1.64(10^{-2})$
0.805	$9.34(10^{-3})$	$1.15(10^{-2})$	$1.10(10^{-2})$	$1.36(10^{-2})$	$1.38(10^{-2})$	$1.85(10^{-2})$	$1.67(10^{-2})$
0.815	$1.34(10^{-2})$	$1.85(10^{-2})$	$1.47(10^{-2})$	$2.10(10^{-2})$	$2.38(10^{-2})$	$2.54(10^{-2})$	$2.65(10^{-2})$
0.825	$1.80(10^{-2})$	$3.11(10^{-2})$	$1.99(10^{-2})$	$3.43(10^{-2})$	$2.54(10^{-2})$	$3.21(10^{-2})$	$4.28(10^{-2})$
0.835	$2.39(10^{-2})$	$3.06(10^{-2})$	$3.19(10^{-2})$	$3.53(10^{-2})$	$3.71(10^{-2})$	$3.40(10^{-2})$	$4.93(10^{-2})$
0.845	$3.14(10^{-2})$	$4.68(10^{-2})$	$4.66(10^{-2})$	$4.06(10^{-2})$	$4.23(10^{-2})$	$5.59(10^{-2})$	$6.23(10^{-2})$
0.855	$4.03(10^{-2})$	$5.03(10^{-2})$	$5.59(10^{-2})$	$5.92(10^{-2})$	$6.48(10^{-2})$	$6.56(10^{-2})$	$8.64(10^{-2})$
0.865	$5.10(10^{-2})$	$6.39(10^{-2})$	$7.77(10^{-2})$	$7.15(10^{-2})$	$8.34(10^{-2})$	$8.97(10^{-2})$	$9.08(10^{-2})$
0.875	$6.40(10^{-2})$	$7.24(10^{-2})$	0.100	$9.67(10^{-2})$	$9.78(10^{-2})$	0.108	0.126
0.885	$7.98(10^{-2})$	0.101	0.110	0.113	0.118	0.122	0.156
0.895	$9.93(10^{-2})$	0.132	0.135	0.127	0.149	0.157	0.179
0.905	0.1236	0.149	0.154	0.165	0.170	0.183	0.233
0.915	0.1464	0.165	0.181	0.216	0.195	0.242	0.258
0.925	0.1646	0.186	0.207	0.222	0.226	0.246	0.294
0.935	0.1868	0.239	0.259	0.236	0.263	0.280	0.308
0.945	0.2128	0.238	0.272	0.302	0.299	0.306	0.340
0.955	0.2401	0.278	0.311	0.314	0.311	0.363	0.370
0.965	0.2703	0.326	0.353	0.348	0.354	0.411	0.420
0.975	0.3043	0.343	0.407	0.384	0.397	0.443	0.464
0.985	0.3434	0.427	0.411	0.431	0.463	0.498	0.524
0.995	0.3892	0.451	0.457	0.484	0.512	0.522	0.604
1.005	0.4429	0.505	0.536	0.565	0.569	0.588	0.653

1.015	0.4873	0.584	0.619	0.599	0.617	0.635	0.681
1.025	0.5174	0.623	0.610	0.619	0.655	0.704	0.786
1.035	0.5537	0.630	0.661	0.649	0.676	0.750	0.798
1.045	0.5951	0.672	0.662	0.702	0.731	0.745	0.815
1.055	0.6346	0.713	0.720	0.761	0.775	0.801	0.875
1.065	0.6748	0.768	0.776	0.744	0.819	0.827	0.895
1.075	0.7162	0.799	0.824	0.828	0.826	0.881	0.927
1.085	0.7591	0.830	0.835	0.861	0.866	0.935	1.018
1.095	0.8041	0.885	0.944	0.931	0.956	0.957	1.023
1.105	0.8512	0.900	0.934	0.930	0.999	1.031	1.072
1.115	0.8915	0.939	0.959	0.984	0.998	1.080	1.081
1.125	0.9236	1.008	1.033	0.977	1.057	1.086	1.096
1.135	0.9575	1.046	1.065	1.056	1.081	1.111	1.171
1.145	0.9925	1.013	1.093	1.082	1.095	1.101	1.178
1.155	1.0255	1.102	1.080	1.089	1.100	1.157	1.176
1.165	1.0577	1.104	1.078	1.108	1.171	1.141	1.218
1.175	1.0893	1.096	1.091	1.098	1.171	1.178	1.227
1.185	1.1205	1.133	1.161	1.183	1.202	1.184	1.271
1.195	1.1516	1.168	1.196	1.191	1.226	1.258	1.260
1.205	1.1826	1.208	1.239	1.236	1.239	1.254	1.310
1.215	1.2088	1.199	1.230	1.249	1.227	1.283	1.355
1.225	1.2297	1.220	1.245	1.263	1.263	1.317	1.328
1.235	1.2507	1.264	1.273	1.263	1.306	1.289	1.332
1.245	1.2714	1.283	1.310	1.275	1.282	1.310	1.337
1.255	1.2903	1.279	1.290	1.329	1.289	1.289	1.320
1.265	1.3079	1.291	1.272	1.308	1.332	1.309	1.341
1.275	1.3246	1.293	1.313	1.305	1.296	1.308	1.346
1.285	1.3404	1.285	1.326	1.348	1.312	1.326	1.324
1.295	1.3555	1.346	1.341	1.341	1.288	1.332	1.336
1.305	1.3699	1.326	1.320	1.305	1.328	1.328	1.343
1.315	1.3813	1.282	1.278	1.341	1.331	1.354	1.348
1.325	1.3896	1.339	1.337	1.314	1.358	1.316	1.347
1.335	1.3975	1.325	1.311	1.301	1.347	1.339	1.342
1.345	1.4048	1.335	1.335	1.314	1.340	1.340	1.334
1.355	1.4108	1.292	1.305	1.354	1.337	1.296	1.325
1.365	1.4159	1.343	1.334	1.295	1.331	1.311	1.322
1.375	1.4202	1.322	1.355	1.342	1.321	1.313	1.280
1.385	1.4239	1.370	1.334	1.326	1.325	1.296	1.300
1.395	1.4269	1.368	1.383	1.358	1.317	1.310	1.300

(*continued*)

Table IV *(continued)*

r/σ	$W_2(r/\sigma)$	$g(r/\sigma)$					
		$\rho\sigma^3 = 0.175$	$\rho\sigma^3 = 0.200$	$\rho\sigma^3 = 0.225$	$\rho\sigma^3 = 0.25$	$\rho\sigma^3 = 0.30$	$\rho\sigma^3 = 0.35$
1.405	1.4294	1.317	1.325	1.345	1.291	1.283	1.276
1.415	1.4302	1.315	1.307	1.308	1.323	1.301	1.283
1.425	1.4293	1.354	1.317	1.301	1.306	1.256	1.268
1.435	1.4281	1.391	1.312	1.295	1.293	1.271	1.235
1.445	1.4265	1.345	1.281	1.303	1.285	1.267	1.267
1.455	1.4242	1.321	1.285	1.278	1.257	1.258	1.228
1.465	1.4214	1.285	1.303	1.282	1.290	1.242	1.206
1.475	1.4181	1.296	1.270	1.249	1.247	1.212	1.195
1.485	1.4148	1.284	1.282	1.251	1.243	1.196	1.206
1.495	1.4107	1.250	1.271	1.235	1.227	1.214	1.178
1.505	1.4065	1.247	1.304	1.243	1.202	1.203	1.184
1.515	1.4015	1.278	1.285	1.251	1.240	1.188	1.170
1.525	1.2957	1.261	1.225	1.206	1.190	1.183	1.144
1.535	1.2898	1.207	1.246	1.232	1.215	1.167	1.140
1.545	1.3838	1.223	1.241	1.208	1.188	1.159	1.120
1.555	1.3774	1.256	1.217	1.190	1.182	1.145	1.129
1.565	1.3710	1.235	1.201	1.206	1.188	1.153	1.123
1.575	1.3643	1.214	1.188	1.191	1.158	1.127	1.112
1.585	1.3577	1.216	1.185	1.171	1.159	1.137	1.089
1.595	1.3509	1.211	1.200	1.151	1.145	1.124	1.080
1.605	1.3442	1.206	1.200	1.143	1.147	1.125	1.074
1.615	1.3771	1.196	1.188	1.144	1.143	1.108	1.073
1.625	1.3296	1.172	1.163	1.146	1.125	1.129	1.080
1.635	1.3222	1.195	1.165	1.140	1.106	1.058	1.061
1.645	1.3149	1.156	1.110	1.133	1.074	1.082	1.057
1.655	1.2076	1.163	1.151	1.117	1.097	1.075	1.039
1.665	1.3003	1.146	1.125	1.118	1.103	1.067	1.040
1.675	1.2931	1.130	1.138	1.119	1.086	1.032	1.012
1.685	1.2859	1.129	1.101	1.100	1.069	1.054	1.008
1.695	1.2789	1.124	1.073	1.118	1.070	1.034	0.993
1.705	1.2720	1.139	1.103	1.097	1.056	1.029	1.000
1.715	1.2650	1.107	1.105	1.063	1.061	1.034	1.006
1.725	1.2578	1.092	1.093	1.079	1.050	1.009	1.000

1.735	1.2509	1.103	1.087	1.058	1.042	0.987	0.991
1.745	1.2441	1.122	1.079	1.061	1.022	1.022	0.985
1.755	1.2373	1.083	1.059	1.049	1.037	1.004	0.978
1.765	1.2308	1.085	1.067	1.022	1.065	1.007	0.965
1.775	1.2243	1.058	1.049	1.023	1.005	0.990	0.965
1.785	1.2180	1.049	1.026	1.047	1.008	0.994	0.958
1.795	1.2119	1.039	1.051	1.018	1.012	0.978	0.972
1.805	1.2060	1.045	1.031	1.019	1.005	0.967	0.953
1.815	1.2000	1.068	1.029	1.018	0.998	0.969	0.942
1.825	1.1941	1.028	1.033	1.004	1.003	0.982	0.948
1.835	1.1883	1.032	1.025	0.999	0.988	0.958	0.943
1.845	1.1827	1.035	1.035	1.000	0.991	0.961	0.939
1.855	1.1773	1.011	1.027	1.008	1.005	0.957	0.935
1.865	1.1719	1.035	1.012	1.003	0.994	0.963	0.938
1.875	1.1668	1.032	1.024	0.998	0.982	0.964	0.932
1.885	1.1618	1.004	1.023	1.006	0.976	0.964	0.944
1.895	1.1570	1.019	1.000	0.988	0.975	0.951	0.932
1.905	1.1523	1.014	1.022	0.980	0.969	0.957	0.934
1.915	1.1477	1.029	0.991	0.970	0.953	0.932	0.933
1.925	1.1431	1.029	0.982	0.989	0.979	0.932	0.928
1.935	1.1387	1.021	1.003	1.005	0.966	0.943	0.941
1.945	1.1344	1.005	0.996	0.988	0.968	0.971	0.939
1.955	1.1303	1.002	0.985	0.998	0.970	0.942	0.933
1.965	1.1262	1.000	0.988	0.992	0.981	0.963	0.949
1.975	1.1224	0.987	0.980	0.974	0.959	0.953	0.935
1.985	1.1186	1.003	0.994	0.985	0.965	0.938	0.927
1.995	1.1150	1.011	0.970	0.963	0.974	0.952	0.930
2.005	1.1115	0.999	0.996	0.968	0.978	0.955	0.930
2.015	1.1081	1.006	0.987	0.972	0.962	0.961	0.936
2.025	1.1047	0.993	0.997	0.996	0.965	0.955	0.935
2.035	1.1015	0.997	0.982	0.975	0.970	0.952	0.949
2.045	1.0984	1.002	0.990	0.975	0.957	0.947	0.946
2.055	1.0954	0.983	0.973	0.973	0.971	0.939	0.944
2.065	1.0924	0.985	0.976	0.975	0.980	0.945	0.961
2.075	1.0896	0.981	0.987	0.969	0.972	0.952	0.932
2.085	1.0869	1.003	0.993	0.976	0.979	0.956	0.952
2.095	1.0843	0.993	0.966	0.979	0.964	0.950	0.955
2.105	1.0817	0.980	0.980	0.979	0.985	0.961	0.952
2.115	1.0793	0.986	0.993	0.981	0.990	0.965	0.951

(continued)

Table IV (*continued*)

r/σ	$W_2(r/\sigma)$	$g(r/\sigma)$					
		$\rho\sigma^3 = 0.175$	$\rho\sigma^3 = 0.20$	$\rho\sigma^3 = 0.225$	$\rho\sigma^3 = 0.250$	$\rho\sigma^3 = 0.30$	$\rho\sigma^3 = 0.35$
2.125	1.0769	0.962	0.986	0.975	0.968	0.968	0.960
2.135	1.0745	0.995	1.003	0.982	0.972	0.976	0.954
2.145	1.0723	0.998	0.955	0.970	0.967	0.970	0.956
2.155	1.0701	0.983	0.981	0.981	0.970	0.968	0.955
2.165	1.0680	0.983	0.981	0.962	0.964	0.968	0.970
2.175	1.0660	1.008	0.989	1.004	0.979	0.954	0.987
2.185	1.0641	1.008	0.979	0.991	0.984	0.972	0.967
2.195	1.0622	0.995	0.981	0.975	0.977	0.968	0.980
2.205	1.0604	0.999	0.974	0.971	0.978	0.972	0.972
2.215	1.0586	0.999	0.992	0.988	0.988	0.958	0.986
2.225	1.0659	0.978	0.989	0.985	0.981	0.983	0.992
2.235	1.0552	0.981	0.988	0.979	0.975	0.985	0.977
2.245	1.0536	0.998	0.980	0.981	0.979	0.969	0.981
2.255	1.0521	0.997	1.002	0.988	0.987	0.987	0.994
2.265	1.0506	0.992	1.004	0.980	0.989	0.981	0.987
2.275	1.0492	1.002	0.984	0.983	0.978	0.991	0.968
2.285	1.0478	1.014	0.992	0.980	0.980	0.989	0.996
2.295	1.0464	1.004	0.985	0.983	0.992	1.014	0.997
2.305	1.0451	1.011	0.998	1.011	0.988	0.994	0.994
2.315	1.0438	0.983	1.010	0.990	0.986	0.989	0.988
2.325	1.0426	1.010	1.011	0.993	1.001	1.001	1.011
2.335	1.0414	0.990	0.993	0.976	0.992	0.979	1.007
2.345	1.0403	0.989	1.005	1.003	0.991	0.989	1.015
2.355	1.0392	1.008	0.992	1.002	1.009	0.999	0.998
2.365	1.0381	0.992	1.004	0.992	1.014	1.002	0.997
2.375	1.0371	0.997	0.996	0.994	1.004	1.004	1.009
2.385	1.0361	1.001	1.003	1.004	1.003	0.990	0.995
2.395	1.0351	1.017	0.987	1.015	1.011	1.010	1.005
2.405	1.0341	1.013	1.013	1.014	1.007	0.994	1.013
2.415	1.0332	0.998	1.029	1.005	1.010	1.005	1.000
2.425	1.0323	1.005	1.006	1.010	1.009	1.011	1.003
2.435	1.0315	1.000	0.999	0.994	1.022	1.008	1.009
2.445	1.0307	1.001	1.027	0.983	1.001	1.012	1.007

2.455	1.0298	1.005	1.004	1.009	1.024	0.999	1.008
2.465	1.0291	1.013	1.014	1.007	1.011	0.999	1.018
2.475	1.0283	1.004	1.008	1.016	1.013	1.010	1.020
2.485	1.0276	0.998	1.006	1.018	1.002	1.017	1.020
2.495	1.0269	1.005	1.013	1.001	1.014	1.014	1.013
2.505	1.0261	1.020	1.011	0.995	1.008	1.005	1.027
2.515	1.0255	1.010	1.016	0.992	1.007	1.013	1.005
2.525	1.0248	1.002	1.018	1.014	1.007	1.020	1.027
2.535	1.0242	1.023	1.016	1.009	1.020	1.021	1.023
2.545	1.0236	0.991	1.011	1.001	1.007	1.013	1.008
2.555	1.0230	0.996	1.021	1.015	1.012	1.007	1.024
2.565	1.0224	1.008	1.017	1.014	1.010	1.014	1.017
2.575	1.0219	1.006	1.010	1.016	1.017	1.009	1.024
2.585	1.0213	1.010	1.009	1.003	1.008	1.013	1.010
2.595	1.0208	1.003	1.012	1.004	1.023	1.015	1.006
2.605	1.0203	1.023	1.029	1.008	1.005	1.019	1.010
2.615	1.0198	1.014	1.005	1.007	1.024	1.020	1.027
2.625	1.0193	1.003	1.029	1.009	0.994	1.014	1.014
2.635	1.0188	0.992	1.008	1.005	1.011	1.014	1.019
2.645	1.0184	1.026	1.007	1.016	1.012	1.001	1.012
2.655	1.0180	1.011	1.016	1.012	0.998	1.014	1.014
2.665	1.0175	1.007	1.005	1.012	1.014	1.007	1.014
2.675	1.0171	1.026	1.016	1.019	1.017	1.008	1.013
2.685	1.0167	1.004	1.009	1.009	1.013	1.032	1.017
2.695	1.0163	1.007	1.028	1.010	1.016	1.017	1.010
2.705	1.0159	1.009	1.020	1.021	1.011	1.016	1.014
2.715	1.0155	1.024	1.016	0.999	1.017	1.011	1.011
2.725	1.0152	1.008	1.003	1.002	1.011	1.000	1.015
2.735	1.0148	1.009	1.005	0.995	1.011	1.012	1.020
2.745	1.0145	1.003	1.025	1.002	1.011	1.017	1.014
2.755	1.0142	1.012	0.995	1.013	1.019	1.017	1.018
2.765	1.0138	1.016	1.003	1.004	1.014	1.009	0.999
2.775	1.0135	1.012	1.035	1.010	1.003	1.005	1.012
2.785	1.0132	1.014	1.002	1.013	1.010	1.010	1.007
2.795	1.0129	1.016	1.006	1.004	0.998	1.010	1.007
2.805	1.0126	1.003	0.997	1.004	1.009	0.999	1.004
2.815	1.0123	1.001	0.997	1.004	1.012	0.997	1.001
2.825	1.0120	0.995	1.013	1.007	1.009	1.005	1.003

(*continued*)

Table IV (*continued*)

r/σ	$W_2(r/\sigma)$	$g(r/\sigma)$					
		$\rho\sigma^3 = 0.175$	$\rho\sigma^3 = 0.20$	$\rho\sigma^3 = 0.225$	$\rho\sigma^3 = 0.250$	$\rho\sigma^3 = 0.30$	$\rho\sigma^3 = 0.35$
2.835	1.0118	1.006	0.999	0.995	1.001	1.000	0.993
2.845	1.0115	1.007	1.016	1.011	1.016	1.009	0.999
2.855	1.0112	0.995	1.016	1.010	0.999	1.007	1.007
2.865	1.0110	1.013	1.028	1.013	1.008	1.003	1.016
2.875	1.0107	0.997	0.996	1.003	1.002	0.998	0.993
2.885	1.0105	1.011	1.010	1.004	1.009	0.996	1.003
2.895	1.0103	1.021	1.001	1.010	1.017	1.002	1.002
2.905	1.0100	1.012	1.018	1.002	0.998	1.011	1.004
2.915	1.0098	1.018	1.003	0.993	0.999	1.012	1.003
2.925	1.0096	0.996	1.006	1.008	1.002	1.006	1.002
2.935	1.0094	0.992	1.006	1.016	0.999	0.999	1.001
2.945	1.0092	1.001	1.015	0.999	1.008	1.006	0.998
2.955	1.0090	1.018	1.005	1.000	1.005	0.995	1.000
2.965	1.0088	0.992	1.004	0.996	1.009	0.993	0.990
2.975	1.0086	1.007	1.007	0.998	0.991	0.997	0.983
2.985	1.0084	1.006	1.012	1.004	0.995	1.001	0.990
2.995	1.0083	0.998	1.009	1.005	1.000	0.993	0.993
3.005	1.0081	1.009	1.002	0.997	1.002	0.994	0.997

Given this tabulated $w(r)$, the analogy previously pointed out with classical statistical mechanics permits us to use the Monte Carlo method to evaluate $g(r)$. For this case, a box of side L with nearest-neighbor periodic boundary conditions was set up to contain $N_s = 128$ particles, the value of L being chosen to make the density of the Monte Carlo system (N_s/L^3) equal to the density of the fluid system. Configurations were then generated with probability proportional to W; after 64,000 equilibrating configurations, averages of $g(r)$ were computed over the next 256,000 configurations for a number of densities. The resulting $g(r)$ are tabulated in Table III for the temperature $T = 7$ K and in Table IV for $T = 8$ K. Thermodynamic properties and results for potentials other than the Lennard-Jones will be published elsewhere (Bruch *et al.*, 1977).

V. Summary and Discussion

Two approaches, very similar in spirit, to the problem of calculating the structure of fluid ^{4}He at zero and nonzero temperatures, respectively, have been discussed here, and their results presented in some detail in Tables I, III, and IV. Concerning these tables, it is evident that they contain the raw data of the Monte Carlo computer experiments and that the results are given to more digits than are really significant. Depending on the purpose to which these results are to be put, some smoothing may be in order. One often useful procedure, which usually produces a smooth curve, is to average the $g(r)$ values in groups of five, i.e., in effect to take $\Delta r = 0.05\sigma$.

It would appear that the basic assumption involved in each of the two approaches described here, namely, that the ground-state wave function [Eq. (2)] and the Slater sum at finite temperatures [Eq. (24)], can be factored into products of pair functions, is a fairly drastic one. Some work has been done to assess the accuracy of this assumption. It seems clear that the validity of the approximation is in both cases density-dependent, and the results might be expected to worsen with increasing density.

Turning first to the Jastrow case, Brandow (1972) has cited nuclear matter calculations by Day (1966) in which the three-body contributions reduce, in a straightforward averaging approximation, simply to the three-body portion of the Jastrow wave function. Brandow argues that this explains why the Jastrow assumption is so successful and suggests that Day's results may provide a means of correcting the Jastrow wave function. Woo (1972) has suggested a specific form of the three-particle wave function, based on an analysis of the weakly interacting Bose gas. Woo and Coldwell (1972) have incorporated some explicit three-particle terms into a preliminary Monte Carlo calculation and have argued that this procedure, if optimized, should lead to good agreement with experiment using the Lennard-Jones potential.

DeMichelis and Reatto (1974), on the other hand, have argued that their Jastrow wave function produces such good agreement with experiment as to make incorporation of three-body effects unnecessary. A definitive resolution of this controversy can be obtained only by further calculation, but two remarks appear to be in order. First, Woo (1972) and Woo and Coldwell (1972) are almost certainly in error in asserting that any improvement in calculational technique, including even in principle an exact one, will produce good agreement with experiment without modifying the potential; the Lennard-Jones potential is simply not accurate. Second, concerning the calculations that must be done to answer these questions, explicit Monte Carlo calculations with three-body terms, as suggested by Woo, are quite costly, possibly prohibitively so, in computer time, as Woo himself has noted. Techniques borrowed from the perturbation theory of classical liquids (Barker and Henderson, 1976) are more likely to be a fruitful method of investigating this problem.

The accuracy of the effective potential approximation has been the subject of some investigation, but not as much. Since the high-temperature limit of the Slater sum is the Boltzmann factor, it is natural to inquire whether the effective potential approximation is a semiclassical expansion, i.e., an expansion in powers of $\hbar^2$. Gibson (1974) has shown that this is not the case and that the effective potential approximation fails to incorporate a three-body term that is of order $\hbar^2$. Gibson (1975) has shown that this causes an error of about 8% in the third virial coefficient of a hard-sphere system and that the corrections are density dependent.

An intriguing possible connection between the two approaches we have discussed is provided by the fact that both the zero temperature wave function [Eq. (15)] and the Slater sums [cf. Eq. (25)] approach one another for small r, i.e., they both approach Eq. (10) with $\alpha/\sigma = 1.05$ or 1.06. This is not surprising, as they are both obtained from what is in effect a similar formalism, but the connection between them, especially their asymptotic behavior, needs to be elucidated.

As a final remark, ^{3}He is a system at least as interesting as ^{4}He and it should be noted that the effective potential approximation has been applied reasonably successfully to the critical point parameters of ^{3}He by Bruch *et al.* (1974). Near the critical point the exchange effects are presumably small, and their neglect is reasonable. But at $T = 0$ they are quite important. One way of incorporating the exchange effects is to multiply the Jastrow wave function by a Slater determinant of plane waves, thereby making it properly antisymmetric. The resulting formulas reduce more or less to those of a fictitious classical interacting gas if a cluster expansion with respect to the antisymmetrization is performed (Wu and Feenberg, 1962; Wu, 1963; Murphy and Nosanow, 1969). The cluster expansion converges fairly rapidly, and good numerical results were obtained by Schiff and Verlet (1967). In order to obtain a Jastrow wave function from solutions to the two-body Schrödinger equation for ^{3}He, some modification to the procedure discussed in Section III would be needed, possibly along the lines suggested by Pandharipande and Bethe (1973).

ACKNOWLEDGEMENTS

This article would not have been completed without the patience of the editors, the collaboration of Professors L. W. Bruch and I. J. McGee on different phases of the work, a grant of machine time from the University of Missouri Computer Network, and the expert typing of Mrs. Edith Hunter.

References

Achter, E. K., and Meyer, L. (1969). *Phys. Rev.* **188**, 291.
Atkins, K. R. (1959). "Liquid Helium." Cambridge Univ. Press, London and New York.
Axilrod, B. M., and Teller, E. (1943). *J. Chem. Phys.* **11**, 299.
Barker, J. A., and Henderson, D. (1976). *Rev. Mod. Phys.* **48**, 587.
Bijl, A. (1940). *Physica (Utrecht)* **7**, 869.
Brandow, B. (1972). *Ann. Phys. (N.Y.)* **74**, 112–249.
Bruch, L. W., and McGee, I. J. (1967). *J. Chem. Phys.* **46**, 2959.
Bruch, L. W., and McGee, I. J. (1970). *J. Chem. Phys.* **52**, 5884.
Bruch, L. W., and McGee, I. J. (1975). *Budapest Int. Conf. Stat. Phys.* (contributed paper).
Bruch, L. W., McGee, I. J., and Watts, R. O. (1974). *Phys. Lett. A* **50**, 315.
Bruch, L. W., McGee, I. J., and Murphy, R. D. (1977). To be published.
Burgmans, A. L. J., Farrar, J. M., and Lee, Y. T. (1976). *J. Chem. Phys.* **64**, 1345.
Campbell, C. E. (1977). *In* "Progress in Liquid Physics" (C. A. Croxton, ed.). Wiley, New York (in press).
Cashion, J. K. (1968). *J. Chem. Phys.* **48**, 94.
Day, B. (1966). *Phys. Rev.* **151**, 826.
DeMichelis, C., and Reatto, L. (1974). *Phys. Lett. A* **50**, 275.
Dingle, R. B. (1949). *Philos. Mag.* [7] **40**, 573.
Farrar, J. M., and Lee, Y. T. (1972). *J. Chem. Phys.* **56**, 5801.
Feenberg, E. (1970). *Am. J. Phys.* **39**, 684.
Gibson, W. G. (1974). *Mol. Phys.* **28**, 793–800.
Gibson, W. G. (1975). *Mol. Phys.* **30**, 13–30.
Hansen, J.-P. (1970). *J. Phys. (Paris)* **31S**, 67.
Hansen, J.-P. (1971). *Phys. Lett. A* **34**, 25.
Hirschfelder, J. O., Curtiss, C. F., and Bird, R. B. (1954). "Molecular Theory of Gases and Liquids," p. 396. Wiley, New York.
Jastrow, R. (1955). *Phys. Rev.* **98**, 1479.
Keller, W. E. (1969). "Helium-3 and Helium-4." Plenum, New York.
Klemm, A. D., and Storer, R. G. (1973). *Aust. J. Phys.* **26**, 43–59.
Leggett, A. J. (1975). *Rev. Mod. Phys.* **47**, 331.
Lim, T. K., Duffy, K., and Damert, W. C. (1977). *Phys. Rev. Lett.* **38**, 341.
Liu, B., and McLean, A. D. (1974). *J. Chem. Phys.* **59**, 4557.
McGee, I. J., and Murphy, R. D. (1972). *J. Phys. C* **5**, L311–314. (Copyright, Institute of Physics; used with permission.)
McGee, I. J., and Murphy, R. D. (1977). To be published.

McMillan, W. L. (1965). *Phys. Rev.* **138**, A442.
Messiah, A. (1961). "Quantum Mechanics," Vol. I, p. 408. North-Holland Publ., Amsterdam.
Münster, A. (1972). "Statistical Thermodynamics," Vol. I, p. 132. Springer-Verlag, Berlin and New York.
Murphy, R. D. (1972). *Phys. Rev. A* **5**, 331.
Murphy, R. D., and Barker, J. A. (1971). *Phys. Rev. A* **3**, 1037.
Murphy, R. D., and McGee, I. J. (1973). *Phys. Lett. A* **45**, 323.
Murphy, R. D., and Nosanow, L. H. (1969). *J. Low Temp. Phys.* **1**, 215–222.
Murphy, R. D., and Watts, R. O. (1970). *J. Low Temp. Phys.* **2**, 507–519.
Muto, Y. (1943). *J. Phys. Soc. Jpn.* **17**, 629.
Pandharipande, V. R., and Bethe, H. A. (1973). *Phys. Rev. C* **7**, 1312.
Percus, J. K., and Yevick, G. J. (1958). *Phys. Rev.* **110**, 1.
Putterman, S. J. (1974). "Superfluid Hydrodynamics." North-Holland Publ., Amsterdam.
Schiff, D., and Verlet, L. (1967). *Phys. Rev.* **160**, 208.
Siska, P. E., Parson, J. M., Schafer, T. P., and Lee, Y. T. (1971). *J. Chem. Phys.* **55**, 5762.
ter Haar, D. (1971). *In* "Physical Chemistry, An Advanced Treatise" (H. Eyring, D. Henderson, and W. Jost, eds.), Vol. VIIIB, pp. 511–538. Academic Press, New York.
van Leeuwen, J. M. J., Groeneveld, J., and deBoer, J. (1959). *Physica (Utrecht)* **25**, 792.
Watts, R. O. (1973). *In* "Specialist Periodical Reports of the Chemical Society: Statistical Mechanics," (K. Singer, ed.), Vol. 1. Chemical Society, London.
Watts, R. O., and McGee, I. J. (1976). "Liquid State Chemical Physics," Wiley, New York.
Wheatley, J. C. (1975). *Rev. Mod. Phys.* **47**, 415.
Wilks, J. (1967). "The Properties of Liquid and Solid Helium." Oxford Univ. Press (Clarendon), London and New York.
Woo, C. W. (1972). *Phys. Rev. Lett.* **28**, 1442–1445.
Woo, C. W., and Coldwell, R. L. (1972). *Phys. Rev. Lett.* **29**, 1062–1064.
Wu, F. Y. (1963). *J. Math. Phys.* **4**, 1438 (1963).
Wu, F. Y., and Feenberg, E. (1962). *Phys. Rev.* **128**, 943.

Proton Transfer Reactions and Kinetics in Water

Frank H. Stillinger
Bell Laboratories
Murray Hill, New Jersey

I. Introduction and Survey

A. Molecular Nature of Pure Water

Undoubtedly the most important solvent in chemistry and chemical technology is water. Furthermore, this substance plays an active and indispensable role in biology. Chemical reactions that take place in water exhibit enormous diversity and frequently possess great industrial, geological, and medical significance. Under these circumstances, then, it is natural to devote serious attention to specific study of chemical reactions in water at the molecular level.

Among all possible chemical reactions that occur in water, the most fundamental is the reversible dissociation reaction

$$H_2O \underset{k_a}{\overset{k_d}{\rightleftharpoons}} H^+ + OH^-. \tag{1.1}$$

Examination of the theory describing forward and reverse rates, and of the nature of the solvated ions formed, constitutes a major objective in this chapter. In addition, we shall broaden the scope to include other aqueous-medium reactions involving proton transfer.

Basic to the understanding of proton transfer reactions and kinetics in water is an understanding of the molecular nature of water itself. Both experimental and theoretical research on water has recently been intense, and as a result of those activities a reasonably complete picture has begun to emerge (Stillinger, 1975, 1977).

In large measure, the structure and properties of water can be explained by the shape of the individual molecules, and by the tendency of those molecules to hydrogen bond to one another. The ground state of the isolated water molecule displays C_{2v} symmetry; the OH bond lengths are 0.9576 Å, and form angle 104.48° at the oxygen nucleus (Benedict *et al.*, 1956). The hydrogen bonds that hold together the condensed phases of water are in undistorted form essentially linear, with an OH covalent bond of one molecule (the proton donor) pointing toward the back side of a neighbor molecule's oxygen atom (the proton acceptor). Participant molecules in a hydrogen bond normally are little distorted from their isolated geometries. Well-formed hydrogen bonds between water molecules have oxygen–oxygen lengths in the range 2.7–3.0 Å, and strength about 5 kcal/mole (Hankins *et al.*, 1970; Popkie *et al.*, 1973).

Hydrogen bonding between neighboring water molecules achieves its greatest extent in the crystal structure of ordinary hexagonal ice (Fletcher, 1970). There, each molecule participates in precisely four hydrogen bonds to nearest neighbors 2.76 Å away. Toward two of these neighbors the central molecule acts as proton donor, while from the other two it accepts protons. The spatial arrangement of these four bonds is the same as four lines radiating outward from the center of a regular tetrahedron toward its vertices. The existence of this tetrahedral coordination with 109.45° between successive bonds is doubtless encouraged by the fact that the isolated-molecule HOH angle is close to that ideally required for strictly linear hydrogen bonds in the crystal.

The hydrogen bonds in hexagonal ice form a space-filling network in which the bonds are locally arranged in hexagons. These hexagons exist in both "chair" and "boat" conformations, analogous to those adopted by the hydrocarbon cyclohexane (Wiberg, 1964). The low temperature cubic modification of ice is similar, but contains only chair hexagons.

The propensity for water molecules to retain tetrahedral coordination through hydrogen bonds to neighbors, with modest angular deformation, is obvious from examination of the structures of the high pressure ices II–IX (Fletcher, 1970). In each case, space-filling networks of hydrogen bonds exist, but the polygons locally produced include quadrilaterals, pentagons, hexagons, and octagons. Even this rich geometric diversity is further broadened by consideration of water networks in clathrate hydrates; in the case of *tert*-butylamine hydrate the unit cell contains a heptagon of hydrogen bonds (Jeffrey *et al.*, 1967).

Evidently the tendency for water to engage in tetrahedral hydrogen bonding extends beyond the melting point of ice into liquid water. X-Ray and neutron diffraction studies of the liquid strongly suggest this qualitative behavior, but with a significant degree of disruption (Narten, 1972). As would be expected, increasing temperature measurably increases the extent of that disruption (Narten and Levy, 1971).

Computer simulation studies have recently been carried out for liquid water. They include both the Monte Carlo (Lie and Clementi, 1975) and molecular dynamics (Stillinger and Rahman, 1974) techniques. These simulations provide confirmation for conclusions drawn from the diffraction experiments about local tetrahedral order in the liquid. They also provide a wealth of additional geometric information about molecular order and hydrogen bond arrangements that is unavailable from any conceivable experiment (Rahman and Stillinger, 1973).

The conclusion that can be drawn from experimental and theoretical studies is that liquid water consists of a random hydrogen bond network, including strained and broken bonds. Furthermore, hydrogen-bond polygons of all sizes larger than triangles are present, with no marked preference for even or for odd numbers of sides (Rahman and Stillinger, 1973). This random network is labile, with bonds breaking and reforming nearby so as to permit diffusion and fluid flow. There is no compelling evidence to suggest that the random network is broken up into patches or regions unconnected to the remainder, but instead a rather uniform degree of hydrogen bond connectivity obtains throughout.

Compared to most other liquids, water has a high static dielectric constant ε_0. At atmospheric pressure, ε_0 is 87.74 at 0°C, but drops to 55.72 at 100°C (Malmberg and Maryott, 1956). To a large degree these high values reflect the polar nature of the water molecule, which in isolation possesses static dipole moment

$$\mu = 1.855 \times 10^{-18} \text{ esu cm} \tag{1.2}$$

(Dyke and Muenter, 1973). It is a simple matter to show that *locally uncorrelated* dipoles would alone produce a static dielectric constant determined by the formula

$$\frac{(\varepsilon_0 - 1)(2\varepsilon_0 + 1)}{\varepsilon_0} = \frac{4\pi\rho\mu^2}{k_B T}, \tag{1.3}$$

where ρ is the molecular number density, k_B is Boltzmann's constant, and T is the absolute temperature. By inserting the μ value shown in Eq. (1.2), and the appropriate ρ values, one finds that Eq. (1.3) predicts

$$\begin{aligned} \varepsilon_0 &= 19.70 \quad (0°\text{C}), \\ &= 13.98 \quad (100°\text{C}). \end{aligned} \tag{1.4}$$

That the measured ε_0 values are substantially larger stems from the failure of naive formula (1.3) to incorporate three important phenomena.

1. The molecules can polarize in an externally applied electric field, due to both nuclear deformation and polarization of the electronic distribution.

2. Even in the absence of an external field, interactions between neighbor molecules in the random network tend on the average to increase the molecular dipole moment to a value $\bar{\mu}$ substantially exceeding μ.

3. Nearby pairs of molecules tend to have their dipole directions biased toward parallelism. The total local moment near a given molecule is frequently denoted by $g_K \bar{\boldsymbol{\mu}}$, where g_K is the Kirkwood orientational correlation factor (Kirkwood, 1939).

Unfortunately there is no fully satisfactory theory for ε_0 in water that accounts for all of these effects quantitatively. Nevertheless, rough estimates both for $\bar{\mu}$ and for g_K have often been proposed (Eisenberg and Kauzmann, 1969) with

$$\begin{aligned} g_K &\approx 2.6, \\ \bar{\mu} &\approx 2.4 \times 10^{-18} \text{ esu cm} \end{aligned} \tag{1.5}$$

for water around room temperature, and with both quantities exhibiting negative temperature coefficients.

The rate at which the random hydrogen-bond network in water spontaneously restructures itself can be monitored by measurements of self-diffusion constants (Mills, 1973), viscosity (Stokes and Mills, 1965), and dielectric relaxation times (Collie *et al.*, 1948). Each of these exhibits strong temperature dependence. The respective Arrhenius plots are distinctly curved, with inferred energies of activation increasing as T declines (Eisenberg and Kauzmann, 1969). These increases are particularly noticeable if data for supercooled water are included, and they surely reflect the energy required for thermal disruption of the random hydrogen-bond network. It is interesting to note in this regard that as water at 0°C freezes to hexagonal ice, the self-diffusion constant decreases discontinuously by a factor of about 10^{-5}, while the dielectric relaxation time increases by about 10^6.

Even without further knowledge, one would reasonably expect other rate processes in water, involving proton transfer, to display similar temperature effects. We will see how that expectation is in fact realized.

B. Dissociation and Electrical Conduction Measurements

The conventional definition of pH for a solution containing H^+ ions is

$$\text{pH} = -\log_{10} a_{H^+}, \tag{1.6}$$

where a_{H^+} is the activity of those ions referred to the hypothetical standard state of unit molar concentration in pure water. Of course, single ion activities are unmeasurable by the normal procedures of solution physical chemistry, so evaluation of pH would seem to require auxilliary assumptions or theoretical calculations.

Under ordinary temperature and pressure conditions, the extent of dissociation in pure water is sufficiently small that the resulting solution of H^+ and OH^- ions is essentially ideal. In that case a_{H^+} in Eq. (1.6) may validly be replaced by the molar concentration of H^+ ions, C_{H^+}. Under other conditions leading to high degree of dissociation, or if concentrated solutes are present, this replacement is unjustified.

The dissociation constant for water K_w is given by the ratio of rate constants in Eq. (1.1):

$$\begin{aligned} K_w &= k_d/k_a \\ &= a_{H^+}a_{OH^-}/a_{H_2O}, \end{aligned} \tag{1.7}$$

where of course in the last form shown a consistent set of standard states must be used for the activities a_i. A closely related constant K'_w can be measured electrochemically, by observing emf's for cells without liquid junctions. Specifically

$$K'_w = a_{H^+}a_{OH^-}/a'_{H_2O}, \tag{1.8}$$

where a_{H^+} and a_{OH^-} refer to the hypothetical 1 M standard state, and where

a'_{H_2O} refers to the standard state of pure water ($\approx$ 55.5 moles/liter). In view of the small extent of dissociation under the usual conditions, K'_w undergoes an important simplification to

$$K'_w = (C_{H^+})^2, \tag{1.9}$$

from which pH may then be determined.

In the temperature range 0°–60°C, values determined electrochemically for K'_w have been listed by Robinson and Stokes (1959). These results are well represented by the formula

$$-\log_{10} K'_w = \frac{4471.33}{T} - 6.0846 + 0.017053T. \tag{1.10}$$

Table I reports concentrations of H^+ and OH^- ions produced by dissociation in pure water that have been determined through K'_w.

Table I

ION CONCENTRATIONS RESULTING FROM DISSOCIATION OF PURE WATER AT VARIOUS TEMPERATURES[a]

t(°C)	$C_{H^+} = C_{OH^-}$ (10^{-7}moles/l)
0	0.337
5	0.430
10	0.540
15	0.671
20	0.825
25	1.004
30	1.212
35	1.445
40	1.708
45	2.004
50	2.340
55	2.701
60	3.101

[a] From Robinson and Stokes (1959).

By differentiating $\log_{10} K'_w$ with respect to temperature, standard thermodynamic changes in enthalpy, entropy, and heat capacity may be determined for the dissociation process. The currently preferred values at 298°K for these quantities, respectively, are the following (Hepler and Woolley, 1973):

$$\begin{aligned} \Delta H^0 &= 13.34 \text{ kcal/mole}, \\ \Delta S^0 &= -19.31 \text{ cal/mole deg}, \\ \Delta C_p^0 &= -53.5 \text{ cal/mole deg}. \end{aligned} \tag{1.11}$$

The standard-state volume change for the reaction can either be determined from pressure variation of K'_w, or from dilatometric measurements of dilute acid–base neutralizations. The preferred value at 298°K is (Hepler and Wooley, 1973)

$$\Delta V^0 = -22.13 \text{ cm}^3/\text{mole}. \tag{1.12}$$

This result quantitatively summarizes the tendency for increasing pressure to shift equilibrium in Eq. (1.1) toward greater dissociation (Hamann, 1963). It also suggests that significant electrostriction occurs around the solvated H^+ and OH^- ions, although the manner in which the random hydrogen-bond network reorganizes to accommodate that electrostriction must be determined by other means.

The galvanic cell method is not useful for study of dissociation in water under extreme temperature and pressure conditions. Instead, K'_w and the ionic concentrations may be obtained from the specific conductance and the limiting equivalent conductance of water. Tödheide (1972) reviews the present status of these measurements and their implications. It is interesting to note that K'_w reaches 10^{-2} at about 1000°C and 100 kbar, with no diminution of the trend toward even greater dissociation as T and p continue to rise. Indeed a very recent report (Vereschagin, *et al.*, 1975) seems to indicate that pressure of about 1 Mbar may suffice to produce nearly complete ionic dissociation.

The limiting equivalent conductances $\lambda^0_{H^+}$ and $\lambda^0_{OH^-}$ are far larger than the same quantities for any other ions. Table II provides values in the range 0°–100°C, at 1 atm. Included for comparison are results for Na^+ and Cl^- in water, which can be regarded as "typical" monovalent ions.

Table II

LIMITING EQUIVALENT CONDUCTANCES FOR IONS IN WATER[a,b]

t(°C)	$\lambda^0_{H^+}$	$\lambda^0_{OH^-}$	$\lambda^0_{Na^+}$	$\lambda^0_{Cl^-}$	$\eta\lambda^0_{H^+}$	$\eta\lambda^0_{OH^-}$	$\eta\lambda^0_{Na^+}$	$\eta\lambda^0_{Cl^-}$
0	225	105	26.5	41.0	402	188	47.4	73.3
5	250.1		30.3	47.5	379		45.9	72.0
15	300.6	165.9	39.7	61.4	342	189	45.1	69.8
18	315	175.8	42.8	66.0	332	185	45.1	69.5
25	349.8	199.1	50.10	76.35	311	177	44.6	68.0
35	397.0	233.0	61.5	92.2	286	168	44.2	66.3
45	441.4	267.2	73.7	108.9	263	159	43.9	64.9
55	483.1	301.4	86.8	126.4	243	152	43.7	63.7
100	630	450	145	212	178	127	41.0	60.0

[a] From Robinson and Stokes (1959), p. 465.
[b] λ_i^0 in cm²/ohm equiv., η in centipoise.

For ions that move as spheres in a structureless, viscous solvent, the product $\lambda_i^0\eta$ should be constant (η = solvent viscosity). Table II shows this is not quite the case even for "typical" ions. Instead, the product declines somewhat with rising temperature. However, the corresponding decline is significantly greater for OH^- and quite dramatic indeed for H^+. Once again H^+ and OH^- act anomalously.

The explanation offered years ago by Bernal and Fowler (1933) (following Grottius and Hückel) for the anomalous mobilities of H^+ and OH^- is still regarded as essentially correct. It rests upon the existence of connected

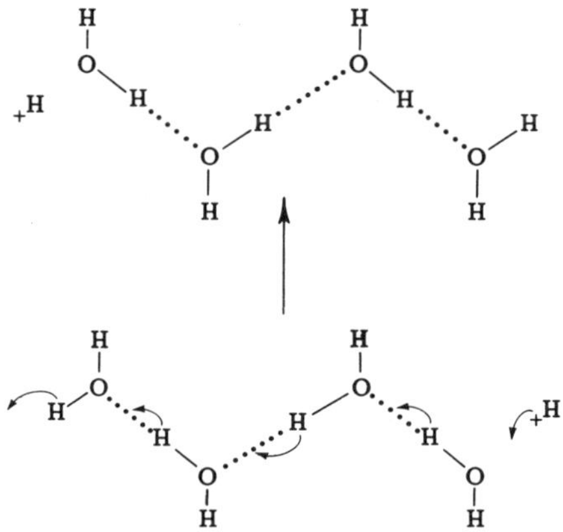

Fig. 1. Electrical conduction of protons in water, by successive hydrogen shifts along a chain of hydrogen bonds.

hydrogen-bond pathways within water and is relevant both to ice and to the liquid. An excess proton at one end of a chain of hydrogen bonds can cause a sequence of proton shifts, each along its own hydrogen bond, whose net result amounts to transfer of the excess proton to the other end of the chain. Similarly, a chain of hydrogen bonds terminating at OH^- (a "missing" proton, or proton "hole") can effectively cause transfer of that ion to its other end by proton shifts in the opposite direction. Figure 1 schematically illustrates the first, and Fig. 2 the second, of these sequential shifts.

Although present knowledge of H^+ and OH^- mobilities in ice is rather imprecise, it appears that freezing water causes both ions to move with greater difficulty. Presently available results for ice (Onsager, 1973) suggest that the H^+ mobility discontinuously drops by a factor of about two, and that of OH^- by roughly a factor of ten, as a result of freezing. At first this seems paradoxical, since hydrogen bond chains are obviously more perfect and more extended in ice than in the liquid. However, one must realize that the rigid ice lattice is largely inconsistent with the negative volume change

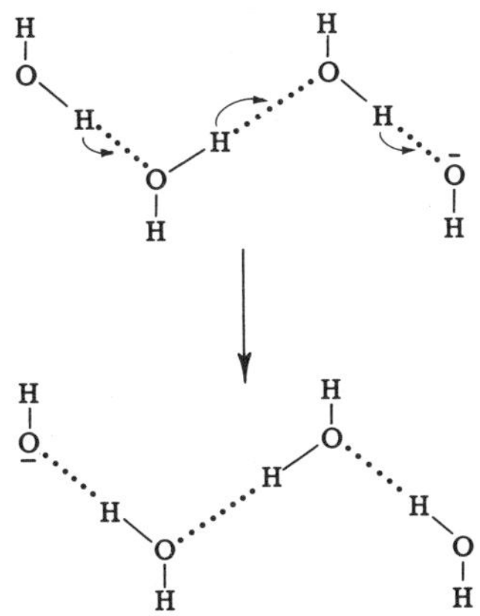

Fig. 2. Mechanism for hydroxide anion mobility in water. The proton "hole" moves by successive hydrogen shifts akin to those shown in Fig. 1.

that normally accompanies dissociation in the liquid. Consequently, the structure around H^+ and OH^- in ice should differ significantly from the corresponding liquid structure, in particular as regards hydrogen-bond lengths. The constraints imposed by the surrounding ice crystal thus appear to retard ionic mobility.

The structural studies of hydrated H^+ and OH^- covered in Sections II,B and C below support this conclusion in a general way, and suggest new details of the ionic transport process. In particular, the thermally induced rate of hydrogen-bond network restructuring must play a fundamental role in explaining the increase in H^+ and OH^- mobilities in the liquid as temperature rises.

C. Isotope Effects

Substitution of deuterium (D) or tritium (T) for hydrogen in water causes important chemical and physical changes. These changes encompass both equilibrium and kinetic properties. Table III provides comparisons between H_2O, D_2O, and T_2O for a short list of properties.

The increase shown in Table III for triple point temperature T_t and maximum-density temperature T_m, as hydrogen isotopic mass increases, clearly indicates an increasing hydrogen bond strength. The same bond-strength phenomenon is also illustrated by decrease in vapor pressure at 20°C, with hydrogen mass increase. However, note that the temperature shifts for T_t are proportionally different than those for T_m, so that the "law of corresponding states" does not apply. Consequently, the isotope effects are more complex than could be explained by mere change in potential energy

Table III

COMPARISON OF PROPERTIES FOR ISOTOPICALLY SUBSTITUTED WATERS

Property	H_2O	D_2O	T_2O
Triple point temperature, T_t(°C)	0.01[a]	3.83[a]	4.49[b]
Maximum-density temperature, T_m(°C)	3.98[a]	11.19[a]	13.40[a]
Critical temperature, T_c(°C)	374.15[a]	370.9[a]	
Static dielectric constant, ε_0 (at 20°C)	80.20[c]	79.89[c]	
Dielectric relaxation time, τ_d (psec, at 20°C)	9.55[a]	12.3[a]	
Self-diffusion constant, D (10^{-5} cm^2/sec, at 25°C)	2.299[d]	1.872[d]	
Dissociation constant, K'_w (10^{-14} moles^2/l^2, at 20°C)	0.681[e]	0.135[f]	
Shear viscosity, η (centipoise, at 30°C)	0.7975[a]	0.969[a]	
Vapor pressure (torr at 20°C)	17.535[g]	15.100[h]	14.553[h]
Second virial coefficient, B (cm^3/mole, at 250°C)	−159.6[i]	−160.7[i]	
Hydrogen ion limiting equivalent conductance, λ_i^0 (cm^2/ohm equiv., at 25°)	349.8[e]	242.4[j]	

[a] Eisenberg and Kauzmann (1969).
[b] Jones (1952).
[c] Vidulich *et al.* (1967).
[d] Mills (1973).
[e] Robinson and Stokes (1959).
[f] Covington *et al.* (1966).
[g] Weast (1975).
[h] Jones (1968).
[i] Kell *et al.* (1968).
[j] Kotowski (1964).

coupling strengths in an otherwise classical statistical mechanical description. Indeed, the critical temperatures T_c (at which most hydrogen bonds may be broken) exhibit a reversed trend.

Quantum-mechanical zero-point motion in water-molecule vibrational modes is important at room temperature owing to the small masses of hydrogen isotopes. Similarly, there are significant quantum effects in rotational motion. The observed differences in thermodynamic properties of H_2O, D_2O, and T_2O stem from the differing extent to which quantum corrections to classical vibration and (hindered) rotation apply to these substances.

The influences of vibrational and rotational quantum corrections to hydrogen bond strength are opposite. For the high-frequency OH stretch modes in water, hydrogen bonding causes a decrease in frequency (Eisenberg and Kauzmann, 1969). The resulting energy reduction upon formation of a hydrogen bond is greater for H than for D, and least for T. Thus if only these stretch-mode vibrational zero-point shifts were at issue, H_2O would tend to have stronger hydrogen bonds than D_2O, and those in T_2O would be weakest of all. The symmetrical bend mode probably shifts in the opposite direction upon bonding (to higher frequency), but it bears relatively little zero-point energy, since it is much the lowest frequency of the three normal modes and so will exert relatively little influence.

Formation of hydrogen bonds turns free rotational motion into librational motion. In ice Ih formed from H_2O these librations fall in the frequency range around 840 cm^{-1} (Eisenberg and Kauzmann, 1969). Conversion of rotation to libration confines angular motion considerably, and must be accompanied by substantial zero-point motion which will *destabilize* the bonded structure. This bond destabilization evidently must be more influential the lighter the hydrogen isotope involved.

In view of the experimental observations, it is clear that librational and bend-mode destabilizations more than compensate for stretch-mode stabilization effects on hydrogen bonding at room temperature.

All rate processes slow down in pure water as hydrogen isotopic mass increases. For the most part these changes are larger in magnitude than those observed for static thermodynamic properties (although K'_w provides an obvious exception). In subsequent sections we shall seek a description of H^+ and OH^- solvation and kinetics which is fully consistent with these facts.

D. Chemical Reactions Involving Proton Transfer

The number of known chemical reactions that involve proton transfer is vast. An exhaustive survey of these reactions would be inappropriate for this article. Instead, it will suffice to illustrate the major classes of proton transfer reactions in water with a few concrete examples, indicating relevant experimental techniques along the way.

The fundamental reaction (1.1) for water itself is known to be characterized by an extremely high association rate constant k_a. Special relaxation methods are required to study such high rates; Eigen and DeMaeyer (1958) employed a rapid electrical pulse technique to study the dissociation field effect (Wien effect for weak electrolytes). They calculated that

$$k_a = (1.4 \pm 0.2) \times 10^{11} \text{ l/mole sec} \tag{1.13}$$

at 25°C. As they themselves stress, this result implies that a sudden mixture of 1 M solutions of strong base and strong acid would be followed by substantially complete neutralization within 10^{-11} sec or so. This extreme speed is consistent only with diffusion-limited rates, i.e., the excess H^+ and OH^- particles need only to find each other to react. No transition state or obvious barrier to reaction exists.

By combining the k_a and K'_w values through Eq. (1.7) (along with the assumption $a_{H_2O} \approx 55.5$ moles/liter), one concludes that

$$k_d = 2.5 \times 10^{-5}/\text{sec} \tag{1.14}$$

at 25°C. This implies that a given intact water molecule will on the average take about 11 hours to dissociate spontaneously.

The strong electric fields that exist in the neighborhood of inorganic ions can severely perturb the dissociation equilibrium of neighboring water molecules. An extreme example is provided by the forced ionization of water molecules comprising the first shell of solvent molecules around an ion. A suitable example would be hydrolysis of the solvated ferric ion:

$$\text{Fe(H}_2\text{O)}_n^{3+} \rightleftharpoons \text{Fe(H}_2\text{O)}_{n-1}\text{(OH)}^{2+} + \text{H}^+. \tag{1.15}$$

It is obvious here that the high positive charge on Fe^{3+} encourages the proton to leave the vicinity, thus effectively reducing the local cationic charge by one unit. The extent of such hydrolysis reactions can be studied effectively by measuring the pH of dilute solutions of the metal salts of interest.

Dissociation of weak acids in water is analogous to reaction (1.1). The homologous series of carboxylic acids (formic, acetic, propionic, butyric, ...) offers a classical family of examples:

$$\text{R} - \text{COOH} \rightleftharpoons \text{R} - \text{COO}^- + \text{H}^+, \tag{1.16}$$

whose dissociation constants (acidity constants)

$$K_a = \frac{[\text{RCOO}^-][\text{H}^+]}{[\text{RCOOH}]} \tag{1.17}$$

can be determined by any convenient pH measurement technique in dilute solution. Members of this series are all stronger acids than water by far, with K_a values clustering around 1.5×10^{-5} mole/liter at 25°C (Fieser and Fieser, 1956) as the organic R-group increases in size. Changes in acidity upon substitution in the alkane chains is an important phenomenon whose explanation resides primarily in modification of the chemical nature of the dissociating molecule itself, rather than in perturbation of the surrounding solvent. As an example, we cite the case of the successively fluorinated acetic acids at 25°C (Chambers, 1973):

$$\begin{aligned}
\text{CH}_3\text{COOH:} &\quad K_a = 1.75 \times 10^{-5} \text{ mole/liter}\\
\text{CH}_2\text{FCOOH:} &\quad K_a = 1.75 \times 10^{-3} \text{ mole/liter}\\
\text{CHF}_2\text{COOH:} &\quad K_a = 3.5 \times 10^{-2} \text{ mole/liter}\\
\text{CF}_3\text{COOH:} &\quad K_a = 1.8 \text{ mole/liter.}
\end{aligned} \tag{1.18}$$

Just as inorganic ions affect the dissociation of solvating water, so too can ionized groups in a large molecule affect dissociation of other ionizable groups in the same molecule. The dicarboxylic acid sequence (oxalic, malonic, succinic, glutaric, ...),

$$HOOC - (CH_2)_n - COOH \qquad (n = 0, 1, 2, 3, \ldots) \tag{1.19}$$

exhibits distinct acidity constants for its first and second proton losses. As n increases, the unfavorable electrostatic interaction between charged ends decreases, so in fact the two ionization constants approach one another.

Conjugate to acid dissociation in water would be association of protons with bases such as primary amines:

$$R - NH_2 + H^+ \rightleftharpoons R - NH_3^+. \tag{1.20}$$

Amino acids offer the possibility of both types of reaction occurring in the same molecule, with formation of "zwitterions" with large dipole moments; in the case of the simplest amino acid, glycine, one finds that the intramolecular proton transfer

$$NH_2 - CH_2 - COOH \rightleftharpoons {}^+NH_3 - CH_2 - COO^- \tag{1.21}$$

is virtually complete at the isoelectric point (pH = 6.064).

A degenerate form of internal proton transfer exists in *o*-aminobenzoic acid, in which the proton of the acid group can geometrically hydrogen-bond to the neighboring nitrogen atom without the necessity for dissociation:

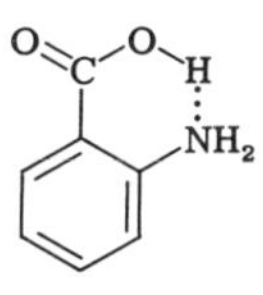

(1.22)

This possibility does not exist for the meta and para forms. Thus the ortho acid is stabilized in its undissociated form, and has the smallest K_a. At 25°C,

$$\begin{aligned} &\text{ortho:} \quad K_a = 1.07 \times 10^{-7} \text{ mole/liter} \\ &\text{meta:} \quad K_a = 1.67 \times 10^{-5} \text{ mole/liter} \\ &\text{para:} \quad K_a = 1.2 \times 10^{-5} \text{ mole/liter.} \end{aligned} \tag{1.23}$$

Dissociation of organic acids can be accompanied by substantial shifts in their electronic excitation spectra. This fact, of course, permits optical spectroscopy to be used to monitor the dissociation degree under various solution conditions. It also leads to the possibility of using strongly absorbing and fluorescing dyes in low concentration as colorimetric pH indicators. Standard indicators are available for use in the pH range 0–14 (Weast, 1975).

Nuclear magnetic resonance (NMR), especially for protons, has also been widely applied to determine acid dissociation in water. The technique has been particularly useful for the study of strong acids such as HCl. In view of the rapidity of proton exchange for such acid solutions, these nuclei give a single resonance signal which is characteristic of the time-averaged environment in which they reside. In these studies, it has been traditional to suppose that dissociation invariably leads to formation of the hydronium ion H_3O^+ (Pople *et al.*, 1959):

$$\mathrm{HA} + \mathrm{H_2O} \rightleftharpoons \mathrm{H_3O^+} + \mathrm{A^-}. \tag{1.24}$$

We shall examine the validity of this solvation assumption later (Sections II and III).

It should also be noted that Raman scattering measurements can be used to observe dissociation of strong acids in concentrated solutions (Young *et al.*, 1959). Discrepancies between NMR and Raman results are occasionally observed and may reflect inadequacy of the proton-solvation hypothesis underlying Eq. (1.24).

In those cases for which proton resonance frequencies in NMR experiments are comparable to their exchange frequencies, lineshapes and positions can be used to determine those exchange rates. Grunwald *et al.* (1957) have exploited this possibility to study the protolysis kinetics of aqueous methylammonium chloride solutions. In the pH range 3–5, they inferred that the proton exchange process

$$\mathrm{CH_3NH_3^+} + \mathrm{CH_3NH_2} \rightleftharpoons \mathrm{CH_3NH_2} + \mathrm{CH_3NH_3^+} \tag{1.25}$$

proceeded through two parallel paths, with roughly equal probability. One path was the direct proton transfer between amines; the other utilized an intervening water molecule or chain of water molecules. Transport of the excess proton thus seems frequently to employ the same basic mechanism that is important in electrochemical transport of protons.

Many chemical reactions in water are specifically catalyzed by hydrogen ions. The hydrolysis of β-propriolactone offers a good example (Long and Purchase, 1950):

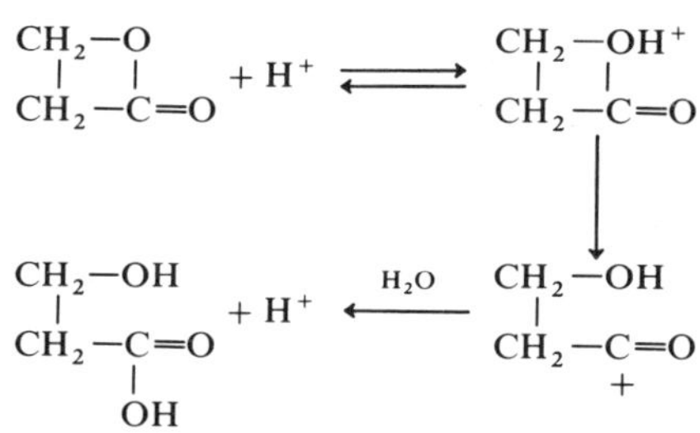

(1.26)

In this case, ring opening precedes addition of the water molecule, perhaps because of ring strain. But with γ-butyrolactone the corresponding proton-catalyzed hydrolysis appears to involve water complexation before ring opening (Long *et al.*, 1951).

Rates of hydrolysis reactions that are specifically catalyzed by hydrogen ions may themselves be used to measure the strengths of acids. Having established the relevant rate constants by prior studies, the degree of dissociation of a given acid in water can be determined by its catalytic power; the stronger the acid, the greater its catalytic action.

Biochemistry provides a large number of important proton transfer reactions. The enzyme carbonic anhydrase, which catalyzes hydration of CO_2 and dehydration of HCO_3^- (or H_2CO_3) offers a prominent example. It contains a zinc atom at its active site, to which a water molecule is strongly bound. It has been proposed that this water molecule and others neighboring it from a chain of hydrogen bonds along which rapid proton transfer occurs in the course of the catalyzed reactions. Finney (1977) has reviewed this and related proposals for the action of other hydrolytic enzymes.

The photochemical action of bacteriorhodopsin involves proton transfer (Stoeckenius, 1976). This purple pigment, found in *Halobacterium halobium*, is a membrane-bound protein. Under illumination, it pumps protons across the membrane to create a potential gradient. The resultant stored energy eventually is used to synthesize the energy-rich molecule ATP that is central to the organism's metabolism.

Finally it should be mentioned that tautomeric proton transfer in nucleotide bases, leading to anomalous base pairing in DNA, has been proposed as a mechanism for spontaneous and induced mutation (Dogonadze *et al.*, 1976).

II. Potential Energy Surfaces

A. Bond Breakage

We now examine in detail the elementary process of dissociation for a single water molecule. Specifically, we inquire into the change in electronic ground state energy of the isolated water molecule as the length r of one of its OH bonds adiabatically elongates to infinity. This process is illustrated in Fig. 3. The bond stretch will be carried out with the other OH bond length and the bond angle held fixed at their equilibrium values (0.9584 Å, 104.45°).

The fragments produced in vacuum by the bond stretch are the H atom and OH radical, rather than H^+ and OH^- ions. The OH radical has equilibrium bond length equal to 0.9699 Å (Herzberg, 1950), so in fact the one produced by the process shown in Fig. 3 is slightly compressed.

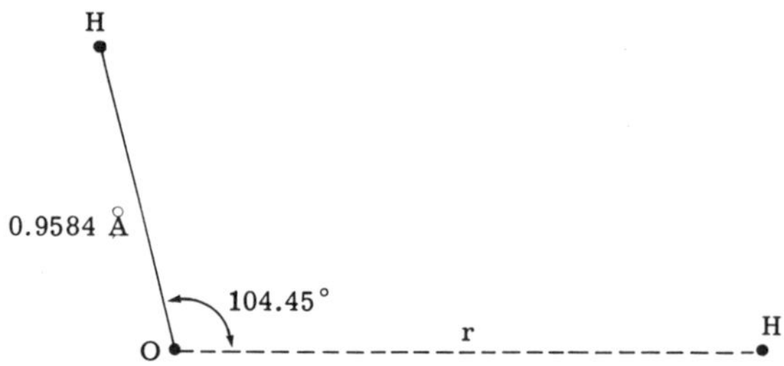

Fig. 3. Water molecule dissociation by single bond stretch. The HOH bond angle and the undeformed bond length retain their equilibrium values, 104.45° and 0.9584 Å, respectively.

Let $E_0(r)$ be the bond stretch energy, which vanishes at $r_e = 0.9584$ Å. For r near r_e an expansion in powers of $r - r_e$ is appropriate:

$$E_0(r) = K_2(r - r_e)^2 + K_3(r - r_e)^3 + K_4(r - r_e)^4 + \cdots. \tag{2.1}$$

There is no reason to doubt that this series has a nonzero radius of convergence. Spectroscopic values are known for the first three coefficients (Smith and Overend, 1972):

$$\begin{aligned} K_2 &= 4.218 \times 10^5 \text{ dyn/cm} \\ K_3 &= -9.57 \times 10^{13} \text{ dyn/cm}^2 \\ K_4 &= 15 \times 10^{21} \text{ dyn/cm}^3. \end{aligned} \tag{2.2}$$

Extraction of higher order terms from either experiment or theory becomes increasingly difficult as the order increases.

As r decreases toward zero, the movable hydrogen nucleus begins to form a "united atom" with the oxygen nucleus. In this limit

$$E_0(r) = (8e^2/r) - C_0 + O(r), \tag{2.3}$$

where the constant C_0 is related to the ground state energy of a hydrogen fluoride molecule with bond length r_e.

The recommended experimental value for the bond dissociation energy in the gas phase process (at 0°K)

$$H_2O \longrightarrow OH + H \tag{2.4}$$

is 118.0 kcal/mole (Darwent, 1970). This energy includes zero-point vibrational contributions from both H_2O and OH. These latter energies are known to be 13.25 kcal/mole for H_2O (Eisenberg and Kauzmann, 1969) and

5.29 kcal/mole for OH (Chamberlain and Roesler, 1955). Combining these values we conclude that

$$\begin{aligned} E_0(\infty) &= 125.96 \text{ kcal/mole} \\ &= 8.7474 \times 10^{-12} \text{ erg/molecule} \\ &= 5.4601 \text{ eV}. \end{aligned} \tag{2.5}$$

Since uncharged fragments are involved, one should expect this limiting value to be approached rapidly with increasing r, probably with exponential order. *Ab initio* quantum-mechanical calculations have not been very successful in attempts to reproduce result (2.5) (Peterson and Pfeiffer, 1972).

The ground electronic state of the water molecule is not the only state that gives unexcited H and OH fragments in the infinite-r limit. Two other excited states of H_2O, a singlet and triplet, will adiabatically produce the same fragments. The corresponding curves of course all lie above $E_0(r)$ for all finite values of r.

The ionization potential of the H atom is 13.530 eV, while the electron affinity of OH is 1.83 ± 0.04 eV (Berry, 1969). Consequently 11.70 ± 0.04 eV (269.9 ± 0.9 kcal/mole) is required to transform the isolated particles H and OH, to H^+ and OH^- by electron transfer. These ionic fragments represent the infinite-r fragments for yet another set of electronic energy curves. From Eq. (2.5) we see that about

$$\begin{aligned} 5.46 + 11.70 &= 17.16 \text{ eV} \\ &= 395.9 \text{ kcal/mole} \end{aligned} \tag{2.6}$$

is the amount of energy that must be expended to produce ionic dissociation of a single water molecule in vacuum. The calorimetric heat of (ionic) dissociation displayed earlier in Eq. (1.11) is far smaller, owing to the very large negative energy of solvating H^+ and OH^- ions in water. Indeed it is this solvation energy that causes dissociation in water to proceed exclusively to ionic products rather than H and OH.

Dipole derivative information (obtained from microwave spectra) helps elucidate the charge redistribution in the water molecule at the beginning of the bond stretch shown in Fig. 3. Clough (1976) provides values for derivatives at the equilibrium geometry that can be resolved into contributions parallel ($\|$) and perpendicular ($\perp$) to the dipole axis of the undeformed molecule:

$$\begin{aligned} (\partial\mu/\partial r)_{\|} &= 0.1568 \times 10^{-10} \text{ esu}, \\ (\partial\mu/\partial r)_{\perp} &= 0.7021 \times 10^{-10} \text{ esu}. \end{aligned} \tag{2.7}$$

One should expect substantially larger values for these effective charges when other water molecules are present to solvate the fragments.

We turn next to an examination of the specific solvation structures produced in water around the ionic H^+ and OH^- dissociation fragments.

B. Hydrated Proton

In order to understand the solvation state of H^+ in liquid water, it is logical first to examine the properties of small hydrate clusters $H^+(H_2O)_n$. These clusters can be generated experimentally in the gas phase, and their thermodynamic properties can be measured (Kebarle *et al.*, 1967). Furthermore, for small n at least, it is possible to carry out reasonably accurate quantum-mechanical calculations to determine structures (Newton and Ehrenson, 1971).

The singly hydrated proton ($n = 1$) is isoelectronic with the ammonia molecule. Therefore it is not surprising that the resulting hydronium ion H_3O^+ is pyramidal, and has C_{3v} symmetry, just as NH_3 does. Kollman and Bender (1973) have carried out *ab initio* quantum-mechanical calculations on the ground electronic state of H_3O^+, with an accuracy claimed to be near the Hartree–Fock limit. The stable C_{3v} structure obtained in their investigation has OH bond lengths equal to 0.963 Å, and HOH angles equal to 112.5°. In comparison with the water molecule geometry, these parameters suggest an enhanced effect of proton–proton repulsion, as might well have been expected in a cationic complex.

Kollman and Bender's study indicates that the inversion barrier (for the $C_{3v} \rightarrow D_{3h} \rightarrow C_{3v}$ transformation) in H_3O^+ lies in the range 2–3 kcal/mole. This is substantially lower than the corresponding barrier height 5.78 kcal/mole in the ammonia molecule (Swalen and Ibers, 1962). Kollman and Bender present energy results calculated for several configurations near the indicated stable point, from which an approximate potential surface could partially be reconstructed.

After accounting for the energy of zero-point nuclear motion, Kollman and Bender estimate from their calculations that the proton affinity of H_2O is 167.5 kcal/mole, and that the deuteron affinity of D_2O (to form D_3O^+) is 170 kcal/mole. The former agrees moderately well with van Raalte and Harrison's value (151 ± 3 kcal/mole) obtained experimentally by gas phase mass spectrometry (van Raalte and Harrison, 1963). The D_2O value is significantly smaller than the estimate 184 ± 7 kcal/mole proposed by DePas *et al.* (1968) also on the basis of mass spectrometry.

The H_3O^+ grouping has been identified in several hydrate crystals. Needless to say, one expects its structure to be subject to significant geometric distortions due to crystal forces. Nevertheless the available crystal data seem to indicate a preference to adhere to the cited pyramidal C_{3v} shape, thereby confirming the conclusions based on quantum-mechanical calculations. Some of the relevant crystal information has been critically discussed by Almlöf and Wahlgren (1973).

The doubly hydrated proton, $H_5O_2^+$, has been examined theoretically by Newton and Ehrenson (1971) and by Kraemer and Diercksen (1970). Both groups utilized approximate Hartree–Fock solutions to the electronic wave equation. These independent studies agree that $H_5O_2^+$ consists of two water molecules (with antiparallel dipole directions and perpendicular molecular planes) connected by a short symmetrical hydrogen bond. The structure has symmetry D_{2d} and is illustrated in Fig. 4. Owing to a difference in the

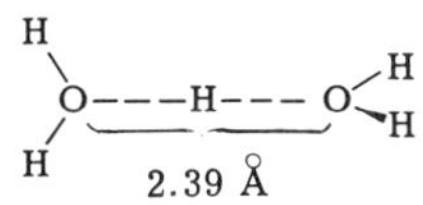

Fig. 4. Stable structure inferred for $H_5O_2^+$ from *ab initio* quantum-mechanical calculations. The terminal water molecules reside in mutually perpendicular planes, both of which contain the central bridging hydrogen.

respective sets of basis functions, the short O–O bond lengths are slightly different in the two studies; Newton and Ehrenson find 2.36 Å, while Kraemer and Diercksen obtain 2.39 Å. There seems to be little doubt that the bridging hydrogen is the one still to be regarded as "the hydrogen ion," since unlike the four other hydrogens it is significantly farther from either oxygen than the normal OH covalent bond length in the water molecule.

The short symmetrical hydrogen bond in $H_5O_2^+$ is closely analogous to the one that exists in the isoelectronic bifluoride anion, FHF^-. At least in the former, Kraemer and Diercksen point out that the bridging proton moves in a very flat potential well along the O–O axis, which no doubt must be relevant to ease of H^+ transport in an extended aqueous medium.

In addition to displacement of the bridging hydrogen, certain other distortions of the $H_5O_2^+$ structure shown in Fig. 4 can occur at relatively low energy cost. In particular, it is easy to rotate one of the terminal H_2O units about its oxygen so as to produce a pyramidal arrangement with the central hydrogen. Thus one readily converts the symmetrical structure shown to an asymmetrical $(H_3O^+)H_2O$, a singly hydrated pyramidal hydronium ion.

The energy of addition of a second water molecule to H_3O^+ to form $H_5O_2^+$ is far less than the energy of the first hydration of H^+ quoted above. Kraemer and Diercksen find this second binding energy to be 32.24 kcal/mole, compared to the experimental value 35.4 kcal/mole (Kebarle *et al.*, 1967).

The crystal structures of several acid hydrates reveal the presence of $H_5O_2^+$. Perchloric acid dihydrate offers a good example (Olovsson, 1968). In this crystal the bridged oxygen–oxygen distance is observed to be 2.424 Å.

Interactions of the $H_5O_2^+$ ions within the solid distort them from the D_{2d} symmetry presumed to obtain for the isolated species. $H_5O_2^+$ ions with different distortions (but similar short O–O distances) occur in the di- and trihydrate crystals of HCl (Lundgren and Olovsson, 1967).

The most systematic quantum-mechanical study of higher hydrates of H^+ has been carried out by Newton and Ehrenson (1971). They were able to perform geometry searches, in varying degrees of completeness, for ions up to $H^+(H_2O)_5$. The complexes studied tended always to exhibit shorter hydrogen bonds than would exist in the corresponding uncharged clusters of water molecules without the extra proton.

In the case of $H^+(H_2O)_3$ (i.e., $H_7O_3^+$) Newton and Ehrenson concluded that the most stable structure is "open," not "cyclic." It can best be described as a flattened H_3O^+, two protons of which form hydrogen bonds to pendant H_2O's. The resulting complex has C_{2v} symmetry, and its two equivalent O–O distances are found to be 2.46 Å. Although subject to the distortive effect of crystal forces, these open $H_7O_3^+$ cations have also been observed in solid hydrates (Almlöf, 1972).

The symmetrically hydrated hydronium ion $H_3O^+(H_2O)_3$ (or $H_9O_4^+$) has been prominently advocated as an important chemical species in liquid water (Eigen and DeMaeyer, 1958). The original suggestions indicated that this species ought to be pyramidal, following the presumed shape of H_3O^+. However, Newton and Ehrenson find that the stable isolated $H_9O_4^+$ has incorporated a flattened H_3O^+ unit to yield symmetry D_{3h}. Figure 5 shows

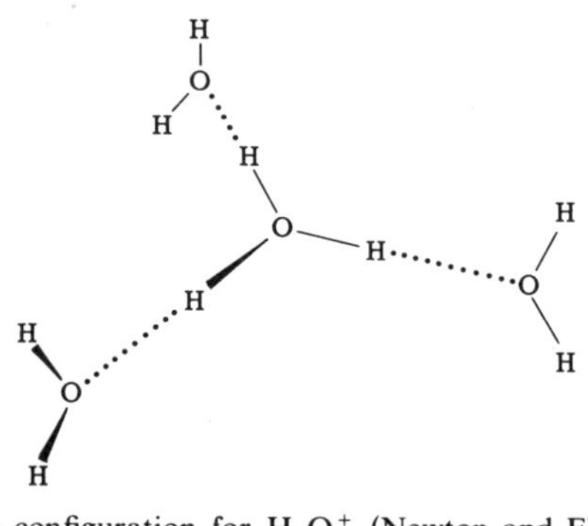

Fig. 5. Lowest energy configuration for $H_9O_4^+$ (Newton and Ehrenson, 1971). The three terminal water molecules are each perpendicular to the plane of the flattened central H_3O^+.

its structure. The three equivalent O–O distances are found to be 2.54 Å. No doubt the flattening present in $H_9O_4^+$ must be due to repulsions between the pendant H_2O's; nevertheless, the energy required to reintroduce pyramidal distortion of significant magnitude is probably small.

The grouping $H_9O_4^+$ is found in crystals of hydrogen bromide tetrahydrate (Lundgren and Olovsson, 1968), along with $H_7O_3^+$. The former complex has O–O distances 2.496 Å, 2.587 Å, and 2.588 Å, and exhibits pyramidal distortion.

Insufficient quantum-mechanical information is available at present to draw any firm conclusions about the minimum-energy structure of $H^+(H_2O)_5$ (i.e., $H_{11}O_5^+$). This cationic unit has not been identified yet in hydrate crystals, although it has been observed in irradiated water vapor (Kebarle *et al.*, 1967).

The largest proton hydrate unit that has thus far been identified in the solid state is $H_{13}O_6^+$. This cation is formed when the cage compound $[(C_9H_{18})_3(NH)_2Cl]^+Cl^-$ is crystallized from hydrochloric acid solution (Bell *et al.*, 1975). The $H_{13}O_9^+$ structure displays a very short central hydrogen bond (2.39 Å between oxygens) along which a bridging proton is presumed to be symmetrically placed. The outer O–O hydrogen bonds are 2.52 Å long. The best description of the cation is that it consists of an $H_5O_2^+$ unit, with each of the four pendant hydrogens bonded to single water molecules. No doubt the $H_{13}O_6^+$ observed in the crystal is considerably distorted from its optimal gas-phase geometry.

Probably the two major conclusions to be drawn from the information cited are (a) the extra proton in water clusters tends to produce substantially shortened hydrogen bonds, and (b) the cationic complexes are easily distorted from their most stable configurations. Extrapolating to the limit of infinite solvation degree ($n \to \infty$), we can expect these two conclusions to apply to excess protons in bulk liquid water. In order for the position of the excess proton to change substantially, the region of shortened hydrogen bonding must likewise shift; these concerted positional shifts are aided by ease of structural distortion induced by presence of the excess proton.

C. Hydrated Hydroxide

We turn now to consideration of the anionic complexes $OH^-(H_2O)_n$. Unfortunately, these seem to be rather difficult to generate in the gas phase, so virtually no experimental information is presently available for these species.

As a point of reference, note that the equilibrium bond length for the isolated OH^- is close to the bond length in H_2O. Janoschek *et al.* (1967) report 0.963 Å for an accurate Hartree–Fock calculation for OH^-.

Once again, the most systematic quantum-mechanical study of the hydrated ions is due to Newton and Ehrenson (1971). Their optimized structure for the monohydrate species $H_3O_2^-$ is illustrated in Fig. 6. This anion is found to be planar, with a relatively short O–O distance, 2.45 Å. However the hydrogen bond is asymmetric, with two equivalent positions for the central proton, separated by 0.23 Å. The double-well potential in which the

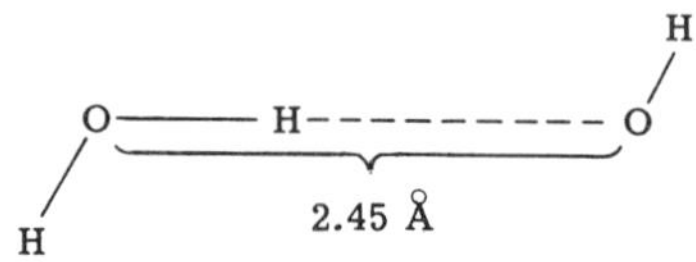

Fig. 6. Optimal structure for $H_3O_2^-$, according to Newton and Ehrenson (1971). The bonding central proton resides at one of the minima in a double-well potential.

proton moves has a predicted barrier height of about 0.13 kcal/mole. The energy of single hydration of OH^- is calculated to be 40.73 kcal/mole.

It is instructive to compare the $H_3O_2^-$ structure with the analogous cation $H_5O_2^+$. Although both incorporate short hydrogen bonds, the anion has apparently a significantly longer one (by 0.09 Å in the Newton–Ehrenson study). This lengthening may explain the existence of a double-minimum potential for the bridging proton in the anion, compared to the very flat potential in the cation. This distinction also may be relevant to the mobility difference between H^+ and OH^- in liquid water that appears in Table II, since it suggests that proton transfers near an excess proton are less frequently retarded by potential barriers than proton transfers near a missing proton ("protonic hole").

The optimized structure found for $H_5O_3^-$ by Newton and Ehrenson is planar, with symmetry C_{2v}. The OH^- acts as proton acceptor for both hydrating water molecules, with equivalent O–O bond lengths of 2.53 Å. Once again this anion length is significantly greater than that found in the same study for the corresponding cation $H_7O_3^+$, namely 2.46 Å.

The trihydrate species $H_7O_4^-$ in the Newton–Ehrenson study achieves greatest stability in a branched structure with each of the water molecules contributing a proton in a hydrogen bond to the central OH^-. The specific configuration of this type studied had C_{3v} symmetry, with the three hydrogen bonds forming a pyramidal shape (with vertex angles for oxygen equal to 110°). The O–O distances were optimized at 2.61 Å, again larger than the 2.54 Å found for the corresponding $H_9O_4^+$ (Fig. 5). It would eventually be desirable to study the change in energy of this branched structure with respect to flattening of the pyramid to see how readily external perturbations could induce such shape changes.

Newton and Ehrenson also carried out preliminary study of the tetrahydrate $H_9O_5^-$ but without sufficient detail and flexibility in configurational options to permit firm conclusions to be drawn.

Overall, the hydrated OH^- complexes seem to involve longer hydrogen bonds and more inflexibility toward rearrangement than the analogous hydrated H^+ complexes. The tendency toward greater bond lengths seems to be supported by crystallographic data, when the previously cited acid hydrates are compared with substances such as sodium hydroxide hydrates (Beurskens and Jeffrey, 1964) and tetramethylammonium hydroxide pentahydrate (McMullan *et al.* 1966).

III. General Theory of Association–Dissociation Reactions

A. Definition of Chemical Species

Results presented in the preceding section suggest that solvated H^+ and OH^- may exist in water with considerable diversity in local structure. Therefore, it may be quite misleading to describe the solution behavior of these ions in terms of discrete chemical entities, such as the frequently mentioned hydronium ion H_3O^+. To avoid unnecessary committment to specific chemical structures, we will now develop a general procedure for description of the dissociation and solvation processes in water. Some elements of this formalism have been advanced previously (Stillinger, 1975).

The central problem is this: Given a set of nuclear positions, $\mathbf{r}_1 \cdots \mathbf{r}_N$ for oxygens, $\mathbf{r}_{N+1} \cdots \mathbf{r}_{3N}$ for hydrogens, attribute all of these nuclei uniquely to H_2O, OH^-, and H^+ species. [We disregard the possible existence of O^{2-}, produced by double dissociation, on physical grounds.] The attribution will be based on the set of oxygen–hydrogen distances produced by $\mathbf{r}_1 \cdots \mathbf{r}_{3N}$.

The only element of arbitrariness to be introduced concerns the choice of a bond dissociation distance L, such that any oxygen–hydrogen pair whose distance exceeds L is never regarded as bonded, either in an H_2O or an OH^- species. A reasonable choice for L seems to be

$$L = 1.375 \text{ Å}. \tag{3.1}$$

This is half the distance between neighboring oxygen nuclei in ice at 0°K and zero pressure and would surely be the natural distance choice for proton assignment to oxygens in that crystal. Note that use of the first three terms in formula (2.1) presented earlier for bond stretching indicates that about 70 kcal/mole are required to increase the normal OH bond in water to length L. Of course this is relevant only for the absence of all other water molecules; their presence would normally reduce the energy needed for bond stretch.

The number of OH pairs of any distance is exactly $2N^2$. We will denote these distances by $l(i, j)$, where i and j are running indices referring, respectively, to oxygens and hydrogens,

$$\begin{aligned} 1 &\le i \le N, \\ N + 1 &\le j \le 3N. \end{aligned} \tag{3.2}$$

Except in the event of zero-probability coincidences, the distances may be ordered in an ascending sequence:

$$l(i_1, j_1) < l(i_2, j_2) < \cdots < l(i_M, j_M), \qquad M = 2N^2. \tag{3.3}$$

Of course each i appears $2N$ times in this sequence, and each j appears N times. The order of appearance depends on the set of positions $\mathbf{r}_1 \cdots \mathbf{r}_{3N}$. In a macroscopic system with a relatively uniform matter distribution, the majority of distances appearing in the sequence (3.3) will have macroscopic magnitudes.

We will be concerned only with the subset of distances obeying the inequality

$$l(i_\alpha, j_\alpha) < L. \tag{3.4}$$

These distances *might* correspond to chemical bonds within the species H_2O and OH^-, but they need not. We shall employ the following "bonding" algorithm for the ordered distance subset obeying (3.4).

1. Bond the two nuclei that provide the minimum distance in the subset list.
2. Remove from the list all distances that involve hydrogens previously bonded and/or involve oxygens bonded twice previously.
3. Return to step 1 if any distances remain in the list.

After this algorithm has exhausted the distance list, the resulting "bonds" will have formed H_2O molecules (2 OH bonds), OH^- ions (1 OH bond), and a certain number of H^+ ions will be left over. Notice that the algorithm does not permit any oxygen to bond to more than two hydrogens, nor does it permit any hydrogen to bond simultaneously to more than a single oxygen. Unbonded H^+ ions may actually be closer to oxygens than the broken bond distance L, but for them to have remained unbonded it is necessary that those oxygens be bonded to pairs of even closer hydrogens.

As time proceeds, the nuclear positions $\mathbf{r}_1 \cdots \mathbf{r}_{3N}$ naturally will change. This will affect OH distances, and consequently the result of the bonding convention will differ. In particular, the number N_w of intact water molecules, of hydroxide anions N_-, and of hydrogen ions N_+ will fluctuate with time t. Of course at all times we have

$$\begin{aligned} N_w(t) &= N - N_+(t), \\ N_-(t) &= N_+(t). \end{aligned} \tag{3.5}$$

There is a variety of ways that bond exchanges can occur (i.e., exchanges of hydrogens between oxygens) without any net change in N_w, N_+, or N_-.

The simplest possibility consists of close approach of H^+ to a water molecule. If this H^+ moves closer to the oxygen atom of the molecule than the larger of its two initial OH bond lengths, then the bonding algorithm requires the new OH distance to be identified as one of the intramolecular bonds. At the same time, the furthest hydrogen of the three loses its bond to the oxygen, and by definition now becomes the unbonded H^+. This transformation is illustrated in Fig. 7.

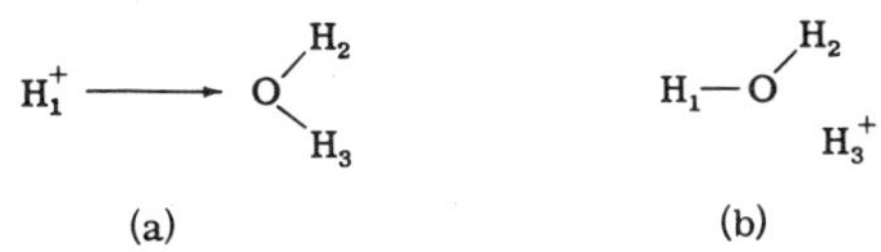

Fig. 7. Bond exchange process for proton approach to a water molecule. (a) The proton at the left is farther from the oxygen than either intramolecular OH bond length. (b) Motion has reduced the initially large distance to a value smaller than one of the original OH bond lengths, thereby requiring re-identification of bonded pairs.

A somewhat more complex bond exchange, involving two water molecules, is shown in Fig. 8. The two molecules possess four internal (intramolecular) bonds; there are in addition four intermolecular OH distances. For any set of bonds assigned by our algorithm, the largest of the first four must be smaller than the smallest of the latter four. If this were not so, the bonding algorithm would have identified the bonds differently. As shown in Fig. 8, a close collision between the two water molecules can require change in bonding, so as to restore compliance with the stated inequality between four intramolecular and four intermolecular distances. As in the preceding simpler case, the new bonding scheme may or may not persist for a substantial length of time, depending on the subsequent motions of the nuclei.

It must also be stressed that under the bonding algorithm, processes exist which involve cooperative bond shifts along a sequence of molecules, which are not themselves required to move to cause these shifts. Figure 9 shows how this can occur; H^+ "attacks" one end of a water molecule chain (along which OH distances increase), and by forcing redrawing of OH bonds, it

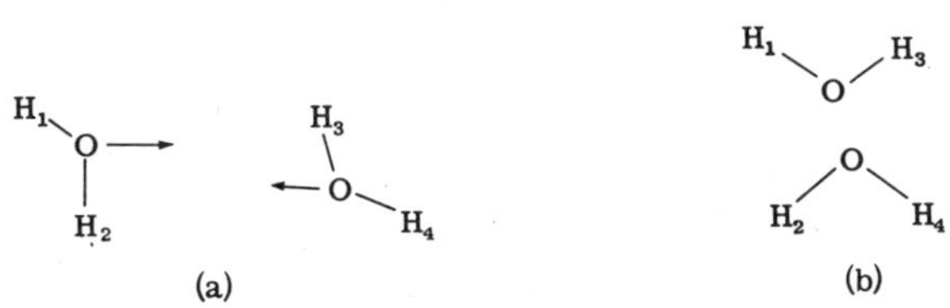

Fig. 8. Simultaneous exchange of hydrogens between two water molecules in a close collision.

succeeds in "freeing" H^+ at the other end. Needless to say, a version of the shift presented in Fig. 9 exists for an excess hydroxide ion, rather than the case shown of an excess proton. One recognizes that the more molecules participating, the rarer will be the configurational circumstances required.

Especially when examining a cooperative shift such as that indicated in Fig. 9, it is important to realize that charge is not discontinuously transported at the instant of rebonding. Of course, electrostatic charge density in

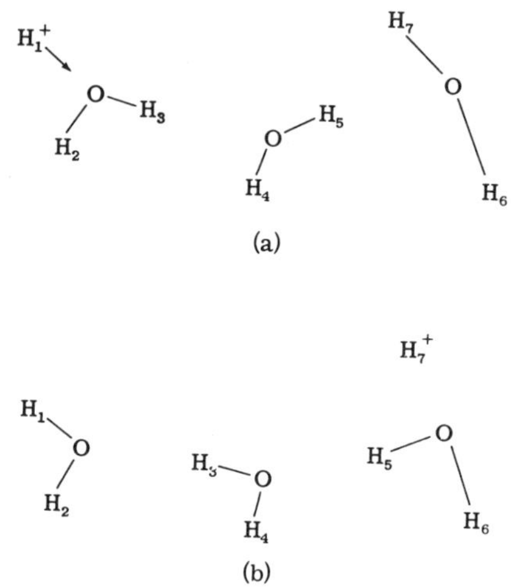

Fig. 9. Cooperative bond transfer along a sequence of molecules, resulting in a different location for the excess proton.

the system is a continuous function of nuclear positions. Under most circumstances we expect excess charge to be localized at or near the ion identified by the bonding algorithm; however, the special molecular arrangements and distortions required by Fig. 9 may tend to delocalize charge over several molecules.

The sequential charge transport illustrated earlier in Fig. 1 stands in distinction to the rate process of Fig. 9. The former requires all intervening protons to move, which in turn causes occurrence of a sequence of simple formal bond shifts.

In addition to the formal exchange processes which conserve N_w, N_+, and N_-, there is a variety of association and dissociation processes encompassed within our bonding scheme which change these species numbers. Just as before, these can be simple, or they can require the cooperative involvement of several intervening molecules. We shall consider these association and dissociation reactions at some length in Section III,C below.

B. Molecular Distribution Functions

Having now introduced an unambiguous identification procedure for intact water molecules and ionic H^+ and OH^- fragments, we next outline a molecular distribution function formalism to describe these species in the liquid phase. The desirability for having such a formalism stems primarily from the inevitable variability in solvation structures around the ions, and from the importance of that variability in affecting kinetics of transport and reaction of those ions. For simplicity of presentation we shall initially use *classical* statistical mechanics. With particles as light as protons it is of course necessary really to use *quantum* statistical mechanics, so at the end of this development the procedure for effecting the necessary modifications will be indicated.

In order to describe the instantaneous configuration of a given water molecule α, we shall use a nine-vector $\mathbf{x}_\alpha$. This comprises full specification of the three nuclear positions. For each of the unbonded protons, only three-vectors are necessary, to be denoted by $\mathbf{s}_\beta$. Six-vectors $\mathbf{t}_\gamma$ will be employed for the hydroxide ions.

The statistical-mechanical theory can be most conveniently developed for the grand ensemble (Hill, 1956). The appropriate generating function is the grand parition function Z_G. We shall initially treat the concentrations of each of H_2O, H^+, and OH^- as independent and therefore we introduce the respective absolute activities y_w, y_+, and y_- as independent parameters. This strategem implies an ability to block attainment of the usual association–dissociation equilibrium between the three species. However, that does not mean we forego knowledge about that equilibrium, because at some later stage we can simply look for a maximum in the grand potential subject to the constraint

$$N_+ = N_-, \tag{3.6}$$

and then identify the concentrations as those appropriate to equilibrium.

The grand partition function has the following form:

$$Z_G = \sum_{N_w, N_+, N_- = 0}^{\infty} \frac{y_w^{N_w} y_+^{N_+} y_-^{N_-}}{(2!)^{N_w} N_w! N_+! N_-!} \times \int d\mathbf{x}_1 \cdots \int d\mathbf{t}_{N_-} \exp[-\beta\Phi(\mathbf{x}_1 \cdots \mathbf{x}_{N_w} | \mathbf{s}_1 \cdots \mathbf{s}_{N_+} | \mathbf{t}_1 \cdots \mathbf{t}_{N_-})],$$

$$\beta = (k_B T)^{-1}. \tag{3.7}$$

Here Φ represents the potential energy (i.e., the ground-electronic state energy hypersurface) for the nuclear configuration specified by $\mathbf{x}_1 \cdots \mathbf{t}_{N_-}$. The denominator factor $(2!)^{N_w}$ in Eq. (3.7) must be inserted to account for proton indistinguishability in each water molecule. Integrations must be restricted to that portion of configuration space which corresponds to bonding of nuclei into the given numbers of molecules and ions, according to the foregoing bonding algorithm.

The connection between Z_G and thermodynamic properties (at least when the system is essentially electrically neutral), is provided by the identification

$$\ln Z_G = \beta p V, \tag{3.8}$$

where p is the pressure and V is the volume. From Eq. (3.7) it is easy to see that the average numbers of each of the three species present can be obtained from activity derivatives of $\ln Z_G$:

$$\langle N_w \rangle = \left(\frac{\partial \ln Z_G}{\partial \ln y_w} \right)_{\beta,\, y_+,\, y_-},$$
$$\langle N_+ \rangle = \left(\frac{\partial \ln Z_G}{\partial \ln y_+} \right)_{\beta,\, y_w,\, y_-},$$
$$\langle N_- \rangle = \left(\frac{\partial \ln Z_G}{\partial \ln y_-} \right)_{\beta,\, y_w,\, y_+}. \tag{3.9}$$

In particular, the functional relation between hydrogen ion concentration and the activity y_+ provided by these identities constitutes a formal way to determine pH in the liquid.

Next we introduce molecular distribution functions $\rho^{(n_w,\, n_+,\, n_-)}$, to describe the occurrence probabilities in the system of sets comprising n_w water molecules, n_+ hydrogen cations, and n_- hydroxide anions, in specified configurations. The formal definition of these quantities in the grand ensemble is as follows:

$$\begin{aligned} \rho^{(n_w,\, n_+,\, n_-)}&(\mathbf{x}_1 \cdots \mathbf{x}_{n_w} | \mathbf{s}_1 \cdots \mathbf{s}_{n_+} | \mathbf{t}_1 \cdots \mathbf{t}_{n_-}) \\ &= Z_G^{-1} \sum_{N_w,\, N_+,\, N_-} \frac{y_w^{N_w} y_+^{N_+} y_-^{N_-}}{(2!)^{N_w}(N_w - n_w)!(N_+ - n_+)!(N_- - n_-)!} \\ &\quad \times \int \cdots \int d\mathbf{x}_{n_w+1} \cdots d\mathbf{x}_{N_w}\, d\mathbf{s}_{n_++1} \cdots d\mathbf{s}_{N_+}\, d\mathbf{t}_{n_-+1} \cdots d\mathbf{t}_{N_-} \\ &\quad \times \exp[-\beta \Phi(\mathbf{x}_1 \cdots \mathbf{t}_{N_-})]. \end{aligned} \tag{3.10}$$

The singlet distribution function $\rho^{(0,\,1,\,0)}(\mathbf{s})$ is equal to the hydrogen ion concentration at position $\mathbf{s}$. In the case of the corresponding hydroxide function $\rho^{(0,\,0,\,1)}(\mathbf{t})$, $\mathbf{t}$ specifies not only position, but orientation and bond length as well, so that this function specifies local density of anions of given orientation and degree of bond stretch. Only after integration over these two internal degrees of freedom will $\rho^{(0,\,0,\,1)}(\mathbf{t})$ yield the local concentration of OH^-. Similarly the singlet water molecule distribution function $\rho^{(1,\,0,\,0)}(\mathbf{x})$

specifies local density of molecules with given position, orientation, and deformation. In liquid water, rotational and translational symmetry applies to the three singlet distribution functions, except possibly near perturbing boundaries.

In the general case of arbitrary (nonnegative) n_w, n_+, and n_-, the molecular distribution function $\rho^{(n_w, n_+, n_-)}$ will reduce to a product of singlet functions if all of the $n_w + n_+ + n_-$ particles are widely separated from one another:

$$\rho^{(n_w, n_+, n_-)}(\mathbf{x}_1 \cdots \mathbf{t}_{n_-}) \to \left[\prod_{i=1}^{n_w} \rho^{(1,0,0)}(\mathbf{x}_i)\right] \times \left[\prod_{j=1}^{n_+} \rho^{(0,1,0)}(\mathbf{s}_j)\right] \times \left[\prod_{k=1}^{n_-} \rho^{(0,0,1)}(\mathbf{t}_k)\right]. \tag{3.11}$$

This asymptotic factorization remains true even if the water is crystalline, for which spatially periodic singlet functions obtain.

The pair molecular distribution functions convey fundamental information about structure in the liquid, and about the nature of H^+ and OH^- solvation. For subsequent purposes it will be most convenient to consider these distribution functions only in the infinite-system-size limit, so that boundary effects do not appear. There exists an obvious set of exact relations, the "local electroneutrality conditions," which must be satisfied by the pair distribution functions. These conditions state that because a partially dissociated liquid is electrically conducting, any charge within the liquid is exactly neutralized (shielded) by an average diffuse accumulation of local charge of opposite sign. In particular this must be true for an H^+ ion, and the resulting identity is

$$\int d\mathbf{s}_2 \left[\frac{\rho^{(0,2,0)}(s_{12})}{\rho^{(0,1,0)}} - \rho^{(0,1,0)}\right] - \int d\mathbf{t}_1 \left[\frac{\rho^{(0,1,1)}(\mathbf{s}_1, \mathbf{t}_1)}{\rho^{(0,1,0)}} - \rho^{(0,0,1)}(\mathbf{t}_1)\right] = -1. \tag{3.12}$$

The corresponding statement for the shielding of an hydroxide ion (with fixed configuration $\mathbf{t}_1$) is

$$\int d\mathbf{s}_1 \left[\frac{\rho^{(0,1,1)}(\mathbf{s}_1, \mathbf{t}_1)}{\rho^{(0,0,1)}(\mathbf{t}_1)} - \rho^{(0,1,0)}\right] - \int d\mathbf{t}_2 \left[\frac{\rho^{(0,0,2)}(\mathbf{t}_1, \mathbf{t}_2)}{\rho^{(0,0,1)}(\mathbf{t}_1)} - \rho^{(0,0,1)}(\mathbf{t}_2)\right] = 1. \tag{3.13}$$

Finally, we have a third relation which states that no net ionic charge will accumulate in the vicinity of a water molecule with fixed configuration $\mathbf{x}_1$, since it is uncharged:

$$\int d\mathbf{s}_1 \left[\frac{\rho^{(1,1,0)}(\mathbf{x}_1, \mathbf{s}_1)}{\rho^{(1,0,0)}(\mathbf{x}_1)} - \rho^{(0,1,0)}\right] - \int d\mathbf{t}_1 \left[\frac{\rho^{(1,0,1)}(\mathbf{x}_1, \mathbf{t}_1)}{\rho^{(1,0,0)}(\mathbf{x}_1)} - \rho^{(0,0,1)}(\mathbf{t}_1)\right] = 0. \quad (3.14)$$

These local electroneutrality conditions can be generalized to provide exact constraints on higher order distribution functions. These generalizations are based on shielding of the total charge of any fixed set of H_2O, H^+, and OH^- particles by the surrounding conductive medium. For present purposes it is unnecessary to dwell upon these generalizations.

In addition to local electroneutrality conditions, the pair distribution functions in a conducting medium are obliged to obey a "second moment condition" (Stillinger and Lovett, 1968). This condition owes its existence to the fact that Coulomb interactions drop off only as inverse distance with increasing separation. We state the second moment condition without proof:

$$\int d\mathbf{r}_{12} r_{12}^2 \left\{\rho^{(0,2,0)}(r_{12}) - 2\int d\boldsymbol{\tau}_2 \rho^{(0,1,1)}(\mathbf{r}_1, \mathbf{t}_2) + \int d\boldsymbol{\tau}_1\, d\boldsymbol{\tau}_2 \rho^{(0,0,2)}(\mathbf{t}_1, \mathbf{t}_2)\right\} = -12C/\kappa^2. \quad (3.15)$$

Here C stands for the common concentration of H^+ and of OH^- ions, and κ is the Debye parameter for the ionic solution:

$$\kappa^2 = 8\pi C e^2/\varepsilon_0 k_B T; \quad (3.16)$$

ε_0 stands for the static dielectric constant of the liquid.

The distance r_{12} appearing in Eq. (3.15) is a distance between two ions. For hydrogen ions there is no ambiguity; this is simply the distance between the hydrogen nuclei. But for hydroxide ions, it is necessary to make a consistent choice of position in that diatomic species for definition of distance. In fact, Eq. (3.15) remains valid whatever convention is chosen (whether it is the O nucleus, or some intermediate point); but in any case the hydroxide six-vector $\mathbf{t}$ must then be resolved into a direct sum of three-vectors:

$$\mathbf{t} = \mathbf{r} \oplus \boldsymbol{\tau}, \quad (3.17)$$

with $\mathbf{r}$ giving the position of the "center" of the hydroxide, and $\boldsymbol{\tau}$ specifying orientation and OH bond length. Note that these latter degrees of freedom are integrated in the second moment condition (3.15).

Owing to the small extent of dissociation in water around room temperature and one atmosphere pressure, the solvated H^+ and OH^- ions constitute a very dilute electrolyte. Consequently, the ionic distribution functions (except at small distances) will accurately be described by the linear Debye–Hückel form (Résibois, 1968):

$$\rho^{(0,2,0)} \sim \rho^{(0,0,2)} \sim C^2[1 - e^2 \exp(-\kappa r_{12})/k_B T \varepsilon_0 r_{12}],$$
$$\rho^{(0,1,1)} \sim C^2[1 + e^2 \exp(-\kappa r_{12})/k_B T \varepsilon_0 r_{12}]. \qquad (3.18)$$

For water at 20°C, the Debye distance $1/\kappa$ is equal to 10,600 Å, which is considerably larger than $C^{-1/3}$, the mean separation between like ions, which equals 2720 Å under the same conditions.

The ion–water and water–water pair distribution functions will approach their long-range limits [described by Eq. (3.11) above] more rapidly than those functions shown in Eq. (3.18); the specific forms involved will exhibit Debye shielding provided the distance is some substantial fraction of $1/\kappa$.

Details about H^+ and OH^- solvation are primarily contained in the pair functions $\rho^{(1,1,0)}$ and $\rho^{(1,0,1)}$. The form of the first of these bears directly on validity of the assumption that solvated protons are predominately present in water as pyramidal hydronium units H_3O^+ (Eigen, 1964). Similarly, the second function would indicate predominance of an analogous pyramidal $OH^-(H_2O)_3$, if such were indeed an apt description of primary hydroxide solvation.

If the primary hydration structure for H^+ in water is principally the pyramidal H_3O^+, then $\rho^{(1,1,0)}$ should be strongly peaked near the configuration shown in Fig. 10a. In particular, this would involve a rather sharp distribution of the $H^+ \cdots O$ distance at about 0.96 Å, and a second

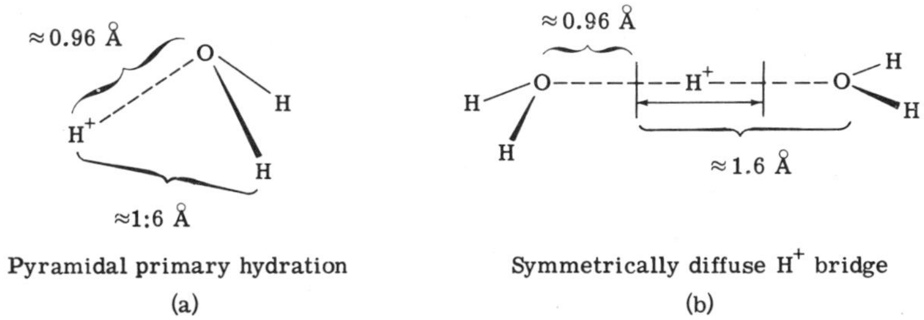

Fig. 10. Primary hydration structures for H^+ in water. The pyramidal "hydronium" in (a) has been advocated by Eigen (1964). Alternative (b) represents a deformable class of proton bridges, with the proton diffusely, but more or less symmetrically, distributed between the participating oxygens.

distinguishable maximum in $H^+ \cdots O$ distances at about 1.6 Å to be consistent with a secondary hydration shell (Eigen, 1964). Furthermore the postulated nonplanarity of the basic H_3O^+ implies that the H^+ is preferentially located about 1.6 Å from the two H's of the water, when this molecule's oxygen is near the cited 0.96 Å distance from H^+.

On the other hand, we have noted in Section II,B that the solvated proton can be (and often is, in crystals) located symmetrically and diffusely along short hydrogen bonds between two essentially equivalent water molecules [Fig. 10b]. If indeed this is the more appropriate description, then $\rho^{(1,1,0)}$ would have to convey a broad single distribution of $H^+ \cdots O$ distances over the range 0.96–1.60 Å, without any marked preference for sharply defined $H^+ \cdots H$ distances.

The water–hydroxide pair distribution function $\rho^{(1,0,1)}$ should possess a strong maximum when the two oxygen atoms are about 2.5 Å apart. Since this represents a strong, short hydrogen bond with the water acting as proton donor, $\rho^{(1,0,1)}$ should achieve its largest value in this distance range in the two equivalent molecular orientations that point covalent OH bonds toward the hydroxide hydrogen. Furthermore, since the hydroxide has no tendency itself to act as proton donor, no significant water molecule probability should be manifest in $\rho^{(1,0,1)}$, along the hydroxide OH bond, at $O \cdots O$ distances in the range 2.5–3.0 Å.

Under those temperature and pressure conditions which lead to small degree of water dissociation, separate partial molar volumes $\bar{V}_+^0$ and $\bar{V}_-^0$ may be attributed to H^+ and to OH^- which reflect the solvation of these individual ions by pure undissociated water. The result (1.12) quoted earlier for the volume of dissociation is essentially composed of the sum of these quantities, both of which are probably negative. Expressions can be written for $\bar{V}_+^0$ and $\bar{V}_-^0$ in terms of singlet and pair distribution functions. For H^+ one derives the following expression:

$$\bar{V}_+^0 = k_B T \kappa_T - \frac{V}{\langle N_w \rangle} \int d\mathbf{x} \left[\frac{\rho^{(1,1,0)}(\mathbf{x}\,|\,\mathbf{s})}{\rho^{(0,1,0)}(\mathbf{s})} - \rho^{(1,0,0)}(\mathbf{x}) \right], \qquad (3.19)$$

wherein κ_T stands for isothermal compressibility of water. An exactly analogous expression gives $\bar{V}_-^0$.

The quantum-mechanical extension of the distribution function formalism requires first that one introduce a suitable complete orthonormal set of eigenfunctions. This complete set is necessary to describe motion of all oxygen and hydrogen nuclei on the ground electronic state hypersurface (the Born–Oppenheimer separation of nuclear and electronic motion should be an accurate description). Then the quantum-mechanical grand partition function has the form of an operator trace:

$$Z_G = \text{Tr} \exp[-\beta(\mathscr{H} - \mu_O \mathscr{N}_O - \mu_H \mathscr{N}_H)]. \qquad (3.20)$$

Here $\mathscr{H}$ is the nuclear-motion Hamiltonian, $\mathscr{N}_O$ and $\mathscr{N}_H$ are number operators for O and H nuclei, respectively, and μ_O and μ_H are the corresponding nuclear chemical potentials.

The bonding algorithm advocated above (Section III,A) partitions multidimensional nuclear configuration space into regions corresponding to the fixed numbers N_w, N_+, N_- of intact molecules, H^+ ions, and OH^- ions. This geometric partitioning effectively defines number operators $\mathscr{N}_w$, $\mathscr{N}_+$, and $\mathscr{N}_-$ for the respective species, so that the averages $\langle N_w \rangle$, $\langle N_+ \rangle$, and $\langle N_- \rangle$ may be obtained quantum-mechanically as suitable traces:

$$\langle N_\alpha \rangle = Z_G^{-1} \operatorname{Tr}\{\mathscr{N}_\alpha \exp[-\beta(\mathscr{H} - \mu_O \mathscr{N}_O - \mu_H \mathscr{N}_H)]\},$$

$$\alpha = \mathrm{w},\ +,\ -,. \qquad (3.21)$$

The quantum-mechanical distribution functions corresponding to the classical quantities defined earlier in Eq. (3.10) may also be represented as operator traces. Alternatively, one can consider the full density matrix for the operator

$$\exp[-\beta(\mathscr{H} - \mu_O \mathscr{N}_O - \mu_H \mathscr{N}_H)]$$

in coordinate representation, and then produce reduced density matrices by integration. Note, however, that our bonding algorithm requires that the integrations be carried out over rather complicated multidimensional regions.

The passage from classical statistical mechanics to quantum statistical mechanics engenders two basic changes in the molecular description of water. The first concerns zero-point motion, primarily of the light hydrogens, which tends to broaden distribution function maxima (i.e., particles will be somewhat delocalized). Needless to say, this effect will be greatest for the lightest isotopes. The other effect is obvious from Table III, namely, that light-hydrogen-mass quantum effects tend to displace the association–dissociation equilibrium toward ionic products. The implication for singlet distribution functions is obvious, and these in turn affect the higher order distribution functions.

C. Linear Response Theory of Reaction for Pure Water

Our next task is derivation of a fluctuation–dissipation theorem for the dissociation process in liquid water. Although autocorrelation functions for chemical reaction rates have previously been deduced (Kutz *et al.*, 1974), the emphasis and details of the present distinct approach seem to reveal the underlying dynamics more clearly than ever before. We shall develop the purely classical theory first, then we indicate how the quantum-mechanical version can be achieved.

For simplicity, suppose that the system contains precisely N oxygen nuclei and $2N$ hydrogen nuclei. Furthermore let $\mathbf{r}$ and $\mathbf{p}$, respectively, comprise all configuration and momentum coordinates for the $3N$ nuclei. The bonding algorithm presented in Section III, A provides a definition for $N_+(\mathbf{r})$, the number of hydrogen ions to be identified as fragments of dissociated water molecules, for any given $\mathbf{r}$. As stated earlier [Eq. (3.5)], N_+ is the number of hydroxides as well, and $N - N_+$ is the number of undissociated molecules.

Denote by $\Omega(N_+)$ that portion of the full $3N$-dimensional $\mathbf{r}$-space which corresponds to the existence of precisely N_+ hydrogen ions in the liquid. The probability $P(N_+)$ that indeed this is exactly the number of H^+ ions present may be expressed in terms of configuration space integrals in the following obvious way:

$$P(N_+) = \left\{\int_{\Omega(N_+)} d\mathbf{r} \exp[-\beta\Phi(\mathbf{r})]\right\} \Big/ \left\{\sum_{N_+=0}^{N} \int_{\Omega(N_+)} d\mathbf{r} \exp[-\beta\Phi(\mathbf{r})]\right\}. \quad (3.22)$$

Owing to the usual behavior of large numbers, this probability can be adequately represented by a normalized Gaussian distribution:

$$P(N_+) = (K/2\pi)^{1/2} \exp[-\tfrac{1}{2}K(N_+ - \langle N_+\rangle)^2]. \quad (3.23)$$

The quadratic fluctuation is then trivial to calculate:

$$\langle (N_+ - \langle N_+\rangle)^2\rangle = 1/K. \quad (3.24)$$

It is convenient now to introduce a mathematical device for manipulating the average number of hydrogen ions. This device takes the form of a weak external potential $U(\mathbf{r})$ which is applied to the system, with the specific form $\xi N_+(\mathbf{r})$. ξ is a small coupling constant. Obviously this external potential is constant across each of the regions $\Omega(N_+)$, but possesses discontinuities of integer multiples of ξ at their boundaries.

In the presence of the external potential, Eq. (3.22) obviously generalizes to the following:

$$P(N_+, \xi) = \left\{\int_{\Omega(N_+)} d\mathbf{r} \exp[-\beta\Phi(\mathbf{r}) - \beta\xi N_+(\mathbf{r})]\right\} \Big/ \left\{\sum_{N_+=0}^{N} \int_{\Omega(N_+)} d\mathbf{r} \exp[-\beta\Phi(\mathbf{r}) - \beta\xi N_+(\mathbf{r})]\right\}. \quad (3.25)$$

The corresponding change in the Gaussian form (3.23) may easily be shown to be

$$P(N_+, \xi) = (K/2\pi)^{1/2} \exp[-\tfrac{1}{2}K(N_+ - \langle N_+\rangle_0 + \xi/K)^2], \quad (3.26)$$

where $\langle N_+\rangle_0$ denotes the average when $\xi = 0$. One immediately sees that the (linear) effect of external potential $U(\mathbf{r})$ on $\langle N_+\rangle$ will be

$$\langle N_+\rangle = \langle N_+\rangle_0 - (\xi/K). \tag{3.27}$$

The quantity K^{-1} is a susceptibility conjugate to external potential strength ξ. Comparing Eqs. (3.24) and (3.27), one reaches a familiar type of conclusion, namely, that large quadratic fluctuations go hand in hand with large susceptibility to external potential.

We note in passing that the dissociation susceptibility K^{-1} may be expressed in terms of the hydrogen-ion pair distribution function $\rho^{(0,2,0)}$. Specifically, one calculates

$$1/K = \langle N_+\rangle + \int d\mathbf{s}_1 \int d\mathbf{s}_2[\rho^{(0,2,0)}(\mathbf{s}_1, \mathbf{s}_2) - \rho^{(0,1,0)}(\mathbf{s}_1)\rho^{(0,1,0)}(\mathbf{s}_2)], \tag{3.28}$$

wherein the distribution functions are those appropriate for the infinite system-size limit.

Equation (3.27) shows how variation in the strength ξ of the external potential $U(\mathbf{r})$ permits one in principle to manipulate the ion concentration in the water sample. If ξ changes from one steady value to another, the degree of dissociation will shift to accommodate that change. The rate at which the system relaxes to the new point of equilibrium is determined by kinetics of the dissociation process and its reverse. We will examine that relaxation process carefully and by doing so will be able to deduce a molecular expression for the chemical reaction rates.

First we need to examine the phenomenological kinetics implied by the fundamental reaction with which this article was opened:

$$H_2O \underset{k_a}{\overset{k_d}{\rightleftharpoons}} H^+ + OH^-. \tag{1.1}$$

The first-order rate equation implied by (1.1) is

$$\frac{d}{dt}\frac{\langle N_+\rangle}{V} = k_d\left(\frac{N - \langle N_+\rangle}{V}\right) - k_a\left(\frac{\langle N_+\rangle}{V}\right)^2. \tag{3.29}$$

Equation (3.27) shows that the final change in hydrogen-ion concentration caused by a sudden small change $\delta\xi$ in external potential will be $-\delta\xi/VK$; kinetic equation (3.29) requires that the approach to this change will be exponential in time. If the cited change $\delta\xi$ occurs at time t_0, the subsequent change in hydrogen ion concentration will be given by

$$\delta\left(\frac{\langle N_+\rangle}{V}\right) = \frac{\delta\xi}{VK}\{\exp[-k_r(t - t_0)] - 1\} \qquad (t \geq t_0). \tag{3.30}$$

In this expression, the relaxation rate k_r is

$$k_r = k_d + 2(\langle N_+\rangle_0/V)k_a. \tag{3.31}$$

The response to any small, but otherwise arbitrary, time-varying ξ may easily be synthesized from exponential lags of type (3.30). By invoking linearity of response, one obtains the general phenomenological result:

$$\frac{d}{dt}\frac{\langle N_+\rangle}{V} = -\frac{k_r}{VK}\int_{-\infty}^{t} ds \exp[-k_r(t-s)]\frac{d\xi(s)}{ds}. \tag{3.32}$$

In studying relaxation processes such as this it is important to observe response to perturbations of different frequencies, so we can set

$$\xi(t) = \xi_0 \exp(i\omega t), \tag{3.33}$$

where ω must have a negative imaginary part to ensure convergence. Insert this $\xi(t)$ into Eq. (3.32) to find

$$\frac{d}{dt}\frac{\langle N_+\rangle}{V} = -\frac{i\omega\xi_0 k_r \exp(i\omega t)}{VK(k_r + i\omega)}. \tag{3.34}$$

This is the phenomenological expression whose comparison with microscopic theory generates a new expression for k_r.

Chronological evolution of the system of $3N$ nuclei may be described by the phase space probability function $f(\mathbf{r}, \mathbf{p}, t)$. This function satisfies the Liouville equation

$$\partial f/\partial t = i\mathscr{L}(t)f; \tag{3.35}$$

$\mathscr{L}$ is the Liouville operator

$$\mathscr{L}(t) = i\sum_{j=1}^{3N}\left[\frac{1}{m_j}\mathbf{p}_j \cdot \nabla_{\mathbf{r}_j} + \mathbf{F}_j(t)\cdot\nabla_{\mathbf{p}_j}\right]. \tag{3.36}$$

The force $\mathbf{F}_j(t)$ acting on nucleus j consists of two parts. The first, $\mathbf{F}_j^{(0)}$, is due simply to internal forces within the system (including wall forces):

$$\mathbf{F}_j^{(0)} = -\nabla_{\mathbf{r}_j}\Phi, \tag{3.37}$$

and has no explicit dependence on time t. The other part, $\mathbf{F}_j^{(1)}(t)$, is the external force on nucleus j due to $U = \xi N_+$:

$$\mathbf{F}_j^{(1)}(t) = -\xi_0 \exp(i\omega t)\nabla_{\mathbf{r}_j} N_+(\mathbf{r}). \tag{3.38}$$

Notice that these latter forces act impulsively; in the $9N$-dimensional configuration space the vector $\nabla N_+(r)$ vanishes except at the boundaries of the region $\Omega(N_+)$, and at these boundaries it has normal orientation from smaller to larger N_+.

We can split the Liouville operator $\mathscr{L}$ into an unperturbed part $\mathscr{L}_0$, and the equilibrium-shifting perturbation $\mathscr{L}_1(t)$:

$$\mathscr{L}(t) = \mathscr{L}_0 + \mathscr{L}_1(t),$$

$$\mathscr{L}_1(t) = i\sum_{j=1}^{3N}\mathbf{F}_j^{(1)}(t)\cdot\nabla_{\mathbf{p}_j}. \tag{3.39}$$

Under the assumption that f was the equilibrium distribution function f_{eq} in the remote past, and that subsequently $\mathscr{L}_1(t)$ was sufficiently weak that only linear response need be considered, then the perturbed distribution function at time t may be displayed in the following form:

$$f(\mathbf{r}, \mathbf{p}, \mathbf{t}) = f_{eq}(\mathbf{r}, \mathbf{p}) - \xi_0 \beta \int_{-\infty}^{t} ds \exp[i(t-s)\mathscr{L}_0] \times \exp(i\omega s) f_{eq}(\mathbf{r}', \mathbf{p}') \sum_{j=1}^{3N} \nabla_{\mathbf{r}_j'} N_+(\mathbf{r}') \cdot (\mathbf{p}_j'/m_j). \tag{3.40}$$

Here $\mathbf{r}'$, $\mathbf{p}'$ stand for phase-space coordinates which unperturbed motion over interval $t - s$ carries into $\mathbf{r}$, $\mathbf{p}$. The quantity

$$J_+(\mathbf{r}, \mathbf{p}) = \sum_{j=1}^{3N} \nabla_{\mathbf{r}_j} N_+(\mathbf{r}) \cdot (\mathbf{p}_j/m) \tag{3.41}$$

represents the total normal current (in the increasing N_+ direction) crossing the Ω boundaries in the $9N$-dimensional configuration space. Hence we can write for the linear approximation:

$$f(\mathbf{r}, \mathbf{p}, t) = f_{eq}(\mathbf{r}, \mathbf{p}) - \xi_0 \beta \int_{-\infty}^{t} ds \exp(i\omega s) J_+(\mathbf{r}', \mathbf{p}') f_{eq}(\mathbf{r}', \mathbf{p}'). \tag{3.42}$$

The expected rate of change of N_+ at time t likewise can be expressed in terms of the current J_+:

$$\frac{d\langle N_+(t)\rangle}{dt} = \int d\mathbf{r} \int d\mathbf{p} f(\mathbf{r}, \mathbf{p}, t) J_+(\mathbf{r}, \mathbf{p}). \tag{3.43}$$

Of course f_{eq} produces no average rate of change for N_+, but the perturbed part of f does. Using expression (3.42) in Eq. (3.43), one finds

$$\frac{d}{dt}\frac{\langle N_+(t)\rangle}{V} = -\frac{\xi_0 \beta}{V} \int d\mathbf{r} \int d\mathbf{p} \int_{-\infty}^{t} ds \exp(i\omega s) \times J_+(\mathbf{r}, \mathbf{p}) J_+(\mathbf{r}', \mathbf{p}') f_{eq}(\mathbf{r}', \mathbf{p}'), \tag{3.44}$$

or more succinctly,

$$\frac{d}{dt}\frac{\langle N_+\rangle}{V} = -\frac{\xi_0 \beta}{V} \exp(i\omega t) \int_{-\infty}^{0} ds \langle J_+(0) J_+(s)\rangle \exp(i\omega s), \tag{3.45}$$

where $\langle \cdots \rangle$ represents an average taken at equilibrium.

At least within the regime of classical dynamics, Eq. (3.45) represents the exact linear response to the arbitrary-frequency perturbation $\xi(t) N_+(\mathbf{r})$ that we have used to disturb the equilibrium. It must now be compared to the phenomenological expression (3.34). Owing to the elementary stochastic assumptions upon which (3.34) is based, we cannot possibly expect the two expressions to agree precisely. In particular, the full molecular dynamics involves detailed high-frequency motions (e.g., vibrations and librations) that play no explicit part in phenomenological chemical kinetics. Nevertheless we can demand agreement between Eqs. (3.34) and (3.45) for small ω. This can be accomplished by expanding both in formal power series in ω, and demanding equality in corresponding orders. In fact it is consistent to make this demand only in the first three orders, and upon doing so one obtains the three relations:

$$\begin{aligned} \int_{-\infty}^{0} ds \, \langle J_+(0) J_+(s) \rangle &= 0, \\ \int_{-\infty}^{0} ds \, s \langle J_+(0) J_+(s) \rangle &= 1/K\beta, \\ \int_{-\infty}^{0} ds \, s^2 \langle J_+(0) J_+(s) \rangle &= -(2/\beta K k_r). \end{aligned} \tag{3.46}$$

Consequently we have a microscopic expression for the chemical equilibrium relaxation time k_r^{-1} strictly in terms of current autocorrelation function integrals:

$$\frac{1}{k_r} = \left[\int_0^{\infty} ds \, s^2 \langle J_+(0) J_+(s) \rangle \right] \Bigg/ \left[2 \int_0^{\infty} ds \, s \langle J_+(0) J_+(s) \rangle \right]. \tag{3.47}$$

Here we have used the fact that the current autocorrelation average must be an even function of s, because of dynamical reversibility. Equation (3.47) is the major result of this section.

It may be of some pedagogic interest to inquire what form the current autocorrelation function would need to be in order to give the phenomenological result (3.34) exactly. In other words,

$$\frac{i\omega k_r}{K(k_r + i\omega)} \equiv \beta \int_{-\infty}^{0} ds \, \langle J_+(0) J_+(s) \rangle \exp(i\omega s). \tag{3.48}$$

The one-sided Fourier transform may easily be inverted to find

$$\langle J_+(0) J_+(s) \rangle = (2k_r/\beta K) \, \delta(s) - (k_r^2/\beta K) \exp(-k_r |s|). \tag{3.49}$$

The meaning of these two terms is very simple. The first, with infinitesimal duration, represents thermally driven "noise" in N_+, associated with stochastic crossing of the $\Omega(N_+)$ boundaries in configuration space. These fluctuations create small deviations from the equilibrium value of N_+ appropriate for the prevailing temperature and pressure conditions. Consequently, relaxation back to equilibrium ensues, which is described by the second term in Eq. (3.49).

The exact autocorrelation function can be expected to differ in several important ways from the "model function" (3.49). Although the existence of discrete jumps in N_+ will necessarily give rise to a delta-function contribution proportional to absolute temperature, the corresponding multiplier is not given correctly by Eq. (3.49). Instead, the correct expression requires an integral over the entire set of $\Omega(N_+)$ boundaries.

Vibrational and librational motions can also be expected to cause deviations from form (3.49) at short (but nonzero) times. These oscillatory mo-

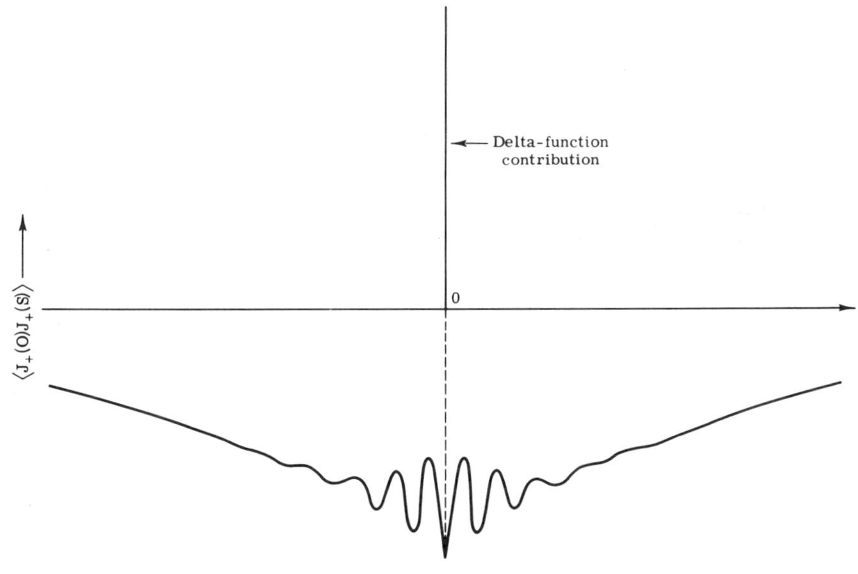

Fig. 11. Current autocorrelation function for the chemical relaxation rate in Eq. (3.47). This is only a schematic representation.

tions will frequently cause the boundaries in configuration space to be crossed back and forth in rapid succession. This will tend to give a strongly negative bias to the autocorrelation at very small times, followed by damped oscillations. The qualitative appearance of $\langle J_+(0)J_+(s)\rangle$ is shown in Fig. 11.

If we were dealing with a slow relaxation process, much slower than vibrational and librational periods in the liquid, then it is plausible that the long-time autocorrelation behavior of J_+ would be nearly a single exponential. However the association–dissociation relaxation in pure water is inherently so fast that such a clean separation of time scales does not always obtain. In particular, one might find important departures from exponential behavior in hot compressed water where the equilibrium ion concentration is high. However, in water at 25°C at 1 atm, the measured relaxation time k_r^{-1} is 3.5×10^{-5} seconds (Eigen and DeMaeyer, 1958), which should indeed lead to nearly exponential behavior over most of the correlation time.

Debye has developed a theory for reaction rates between ionic species in solution (Debye, 1942). With suitable reinterpretation, this theory provides a description of the neutralization reaction between H^+ and OH^- in water, i.e., a description of the processes which determine k_a in Eq. (1.1). The necessary reinterpretation has been advanced by Eigen and DeMaeyer (1958). They conclude that the following expression is valid:

$$k_a = \frac{4\pi e^2}{\varepsilon_0 k_B T}\left[\frac{D_+ + D_-}{\exp(e^2/\varepsilon_0 a k_B T) - 1}\right]. \tag{3.50}$$

Here D_+ and D_- are the apparent self-diffusion constants for the ions, and a is a "reaction distance" such that one assumes that whenever an H^+ and OH^- approach each other this closely they inevitably and quickly react. Using experimental values for k_a, ε_0, and ion mobilities at room temperature, Eigen and DeMaeyer conclude that

$$a = 7.5 \pm 2\ \text{Å}. \tag{3.51}$$

This distance should probably be interpreted as the point of contact between ion solvation regions containing foreshortened hydrogen bonds. Considering the polarization of respective H^+ and OH^- containing regions, such contact would produce a complete chain of such short bonds along which immediate charge transport would occur to produce the neutralization reaction. In this way, Eq. (3.50) provides an important link to the quantum-mechanical (and crystallographic) studies of H^+ and OH^- solvation structures.

This last result is also important in achieving the proper interpretation of the general rate constant expression (3.47). In particular, it suggests that the type of hypersurface crossing illustrated in Fig. 9 (involving formal hopping of H^+ simultaneously over several molecules) may indeed be quite important in water. By dynamical reversibility, the spontaneous dissociation reaction must also involve sudden charge separation by at least distance a, which likewise may frequently proceed by the concerted process of Fig. 9.

Providing that a suitable Hamiltonian model can be constructed to describe dissociating water, the molecular dynamics simulation technique could offer basic insights into the relative contributions to the various types of hypersurface crossings to the reaction rates.

D. Quantum-Mechanical Extension

Just as it was necessary earlier to indicate how molecular distribution functions must be calculated in the quantum-mechanical regime, so too must we examine the corresponding changes for the theory of reaction rates. In a convenient review article, Zwanzig has stated the general relationship between classical and quantum-mechanical expressions for transport and rate constants in current correlation function form (Zwanzig, 1965). This, of course, is the form in which the relaxation rate k_r has been displayed in Eq. (3.47) above. Consequently we can immediately adapt Zwanzig's exposition to the water problem.

The quantum-mechanical extension requires first that the currents J_+ appearing in Eq. (3.47) be replaced by the appropriate operators, and then that the thermal average $\langle \cdots \rangle$ be carried out as a trace with the quantum-mechanical density matrix for the system, $\exp(-\beta \mathscr{H})$.

First note that in the classical regime,

$$J_+(t) = (d/dt)N_+(t); \tag{3.52}$$

$N_+(t)$ is the number of H^+ ions in the system defined by the algorithm stated earlier. We associate with N_+ the quantum-mechanical operator $\mathscr{N}_+$ which, in the position representation, multiplies any given wave function for the system by the appropriate number N_+ for the configuration of interest. The time derivative of $\mathscr{N}_+$ satisfies Heisenberg's equation of motion:

$$\dot{\mathscr{N}}_+ = (i/\hbar)[\mathscr{H}, \mathscr{N}_+], \tag{3.53}$$

where [,] stands for the commutator. The formal solution to Eq. (3.53) is

$$\mathscr{N}_+(t) = \exp(it\mathscr{H}/\hbar)\mathscr{N}_+(0)\exp(-it\mathscr{H}/\hbar). \tag{3.54}$$

In order to display the correct quantum-mechanical generalization, one must use the "Kubo transform" $\dot{\mathscr{N}}_+^{(K)}(t)$ for the operator $\dot{\mathscr{N}}_+(t)$:

$$\dot{\mathscr{N}}_+^{(K)}(t) = \beta^{-1} \int_0^\beta d\lambda \exp(\lambda \mathscr{H})\dot{\mathscr{N}}_+(t)\exp(-\lambda \mathscr{H}). \tag{3.55}$$

We thus have the following quantum version of Eq. (3.47):

$$\frac{1}{k_r} = \left[\int_0^\infty ds\, s^2 \langle \dot{\mathscr{N}}_+^{(K)}(0)\dot{\mathscr{N}}_+(s)\rangle\right] \Big/ \left[\int_0^\infty ds\, s \langle \dot{\mathscr{N}}_+^{(K)}(0)\dot{\mathscr{N}}_+(s)\rangle\right]. \tag{3.56}$$

This last result should be central in any discussion of kinetic isotope effects on the association–dissociation relaxation in water. In some small degree, the isotopic difference in structure between H_2O and D_2O would affect this relaxation rate. But more important probably are (1) the relative rates of network restructuring, and (2) the influence of tunnelling on H^+ or

D^+ motion. The first of these can roughly be correlated with comparison of dielectric relaxation times and self-diffusion constants for H_2O and D_2O (see Table III). The second should be reflected in a diminished *a* value for D_2O compared to H_2O, when the Debye equation (3.50) is used to infer an effective reaction distance.

An obvious need exists at present to measure relaxation rates and ion mobilities accurately in D_2O over an extended density and temperature range, to permit quantitative comparison with the corresponding H_2O results. The comparison should help to assess the magnitude of quantum corrections to kinetic properties in liquid water.

E. Reactions Involving Other Species

The association–dissociation reaction (1.1), of course, is always present in water, even when it acts as a host solvent for other chemical reactions. If these latter reactions themselves involve H^+ or OH^-, there arises a strong coupling between solvent and solute reactions. We shall illustrate the way that theory describes this coupling by examining the specific case of acid dissociation in water. This example serves to clarify concepts required for describing other aqueous reactions of the type mentioned in Section I,D.

The competing reactions to be considered involve four phenomenological rate constants:

$$\begin{aligned} H_2O &\underset{k_a}{\overset{k_d}{\rightleftharpoons}} H^+ + OH^- \\ HA &\underset{K_a}{\overset{K_d}{\rightleftharpoons}} H^+ + A^-. \end{aligned} \tag{3.57}$$

In the dilute solution limit k_a and k_d may, of course, be taken as the rates appropriate for pure water, but in general k_a, k_d, K_a, and K_d will depend upon composition of the binary solution.

First we need to know the relaxation spectrum implied by Eqs. (3.57) for sufficiently small deviations from equilibrium that time-dependent composition variations of $k_a \cdots K_a$ may be neglected. Let $C_0(\nu)$ denote the equilibrium concentration of species ν. Starting from some initial state of disturbed equilibrium, we can represent the time-dependent concentrations of the five species as follows:

$$\begin{aligned} H_2O &: C_0(H_2O) - x(t) \\ HA &: C_0(HA) - y(t) \\ OH^- &: C_0(OH^-) + x(t) \\ A^- &: C_0(A^-) + y(t) \\ H^+ &: C_0(H^+) + x(t) + y(t). \end{aligned} \tag{3.58}$$

The corresponding linear first-order differential equations are easily found to be

$$-\dot{x}(t) = k_d x + k_a[C_0(\mathrm{H}^+) + C_0(\mathrm{OH}^-)]x + k_a C_0(\mathrm{OH}^-)y,$$
$$-\dot{y}(t) = K_d y + K_a[C_0(\mathrm{H}^+) + C_0(\mathrm{A}^-)]y + K_a C_0(\mathrm{A}^-)x. \tag{3.59}$$

The general solution to Eqs. (3.59) consists of linear combinations of two exponentially decaying function pairs of the type

$$x(t) = \exp(-\alpha t),$$
$$y(t) = y(0)\exp(-\alpha t). \tag{3.60}$$

Substituting this form into Eqs. (3.59) yields

$$y(0) = \frac{K_a C_0(\mathrm{A}^-)}{\alpha - S}, \tag{3.61}$$

and

$$\alpha^2 - (s + S)\alpha + sS - k_a K_a C_0(\mathrm{OH}^-)C_0(\mathrm{A}^-) = 0, \tag{3.62}$$

where

$$s = k_d + k_a[C_0(\mathrm{H}^+) + C_0(\mathrm{OH}^-)],$$
$$S = K_d + K_a[C_0(\mathrm{H}^+) + C_0(\mathrm{A}^-)]. \tag{3.63}$$

Were it not for the last term in the left member of Eq. (3.62), the roots for the decay constant α would be precisely s and S. However, this last term, since it is negative, causes the upper root to rise above $\max(s, S)$ and the lower root to sink below $\min(s, S)$.

$$\alpha_\pm = \tfrac{1}{2}\{s + S \pm [(s - S)^2 + 4k_a K_a C_0(\mathrm{OH}^-)C_0(\mathrm{A}^-)]^{1/2}\}. \tag{3.64}$$

Although conditions in the binary solution may be such that $s = S$, notice that the two roots α_+ and α_- always remain distinct. For the larger root, $y(0)$ is positive, and this more rapid of the two relaxation modes has the water and the acid associating (or dissociating) together. The slower relaxation mode has negative $y(0)$, so one of the molecular constituents associates while the other dissociates.

In order to establish contact between the phenomenological equations (3.57) and the microscopic theory, we must once again articulate a procedure for dividing configuration space into regions that correspond to fixed numbers of the five chemical species. Obviously, we want the procedure to reduce to that already advocated for pure water when the constituent HA has zero concentration. Indeed, we shall follow the same bonding algorithm as before, after introducing a suitable prescription for formation of the H–A molecular bond.

In general, the geometric criterion of when an A^- particle and an available H^+ ought properly to be considered bonded should rest upon the nature of the relevant potential energy surface. If A^- were monoatomic (i.e., a halide anion), the spherical symmetry demands that a critical bond formation radius L' be identified [analogous to L for water, Eq. (3.1)]. This radius might best be defined as the outer surface at which the adiabatic H–A potential energy displays some fixed fraction, say one-half, of the maximum possible bond energy. For polyatomic anions, of course, the resulting binding criterion would have to be more complicated. Thus for a long-chain fatty acid, the criterion must assure that the bound proton resides in an appropriate region at the carboxylate end of the molecule. Tautomeric substances such as acetylacetone ($CH_3COCH_2COCH_3$) will in fact exhibit disconnected regions for H^+ attachment.

Assume now that the bonding criterion for H–A has been established. The algorithm for identifying H_2O, HA, OH^-, A^-, and H^+ particles in the solution, given the full set of nuclear coordinates, must proceed as follows.

1. Form a list of all distances between hydrogen nuclei and base-atom nuclei (O for water, and whatever atom or atoms act as points of H^+ attachment in HA). Arrange these distances in ascending order. Coincidental equalities have vanishing probability and may be disregarded.
2. Eliminate all distances that exceed the maximum permissible bond lengths either for water (L) or HA(L').
3. Bond the two nuclei that provide the minimum distance in the remaining list (in the case of HA we additionally may demand that angular requirements be met as well).
4. Remove from the list all distances that involve hydrogens previously bonded and/or involve base atoms bonded the maximum permissible number of times previously (twice for water oxygens, once for base atoms in A^-).
5. Proceed to the next larger distance remaining in the list, and bond the two nuclei involved if appropriate.

By processing the entire list of distances this way the hydrogens are at least partially assigned to the oxygen and A^- particles so as to identify uniquely H_2O and HA molecules, and OH^-, A^-, and H^+ ions. Notice that no hydrogen can simultaneously serve as part of more than just a single H_2O, HA, or OH^- entity. [This procedure consistently neglects hydrogens that may be part of the stable (nonexchangeable) structure of A^-; in the case of some tautomeric substances this simplifying assumption may require revision.]

The binary aqueous solution under consideration may be regarded as having originally been prepared from a fixed number N_I of undissociated water molecules mixed together with N_{II} originally undissociated HA molecules. After mixing, there will be fluctuating numbers $N_w(t)$, $N_{HA}(t)$,

$N_{OH}(t)$, $N_A(t)$, and $N_+(t)$ of the species H_2O, HA, OH^-, A^-, and H^+, respectively. The fluctuations represent only two statistical degrees of freedom, since we have the conservation conditions

$$\begin{aligned} N_w(t) + N_{OH}(t) &= N_I, \\ N_{HA}(t) + N_A(t) &= N_{II}, \\ N_{OH}(t) + N_A(t) &= N_+(t). \end{aligned} \tag{3.65}$$

The anion numbers N_{OH} and N_A offer a convenient pair of independent variables. These discrete variables completely partition the multidimensional configuration space for the system into distinct regions $\Omega(N_{OH}, N_A)$ that yield the indicated numbers of particles through the stated bonding algorithm.

Next we must calculate the linear response of the system to application of a weak external perturbation potential of the type

$$U(\mathbf{r}, t) = [\xi_0 N_{OH}(\mathbf{r}) + \xi_1 N_A(\mathbf{r})] \exp(i\omega t). \tag{3.66}$$

Here ξ_0 and ξ_1 are small constants, possibly complex. The frequency ω should be given a small negative imaginary part to ensure convergence. Obviously U is constant across each region $\Omega(N_{OH}, N_A)$.

As was the case with pure water, the linear response may be computed from the Liouville equation, assuming that classical mechanics applies. This response may be expressed in terms of correlation functions of currents:

$$J_{OH} = dN_{OH}/dt, \qquad J_A = dN_A/dt,$$

which measure crossings of $\Omega(N_{OH}, N_A)$ boundaries. There is no need to reproduce the derivation, which merely parallels that of the earlier case. One obtains the following results:

$$\begin{aligned} \frac{d}{dt}\frac{\langle N_{OH}(t)\rangle}{V} &= -\frac{\beta \exp(i\omega t)}{V}[\xi_0 \Phi_1(\omega) + \xi_1 \Phi_3(\omega)], \\ \frac{d}{dt}\frac{\langle N_A(t)\rangle}{V} &= -\frac{\beta \exp(i\omega t)}{V}[\xi_1 \Phi_2(\omega) + \xi_0 \Phi_3(\omega)], \end{aligned} \tag{3.67}$$

where

$$\begin{aligned} \Phi_1(\omega) &= \int_0^\infty ds \exp(-i\omega s)\langle J_{OH}(0) J_{OH}(s)\rangle, \\ \Phi_2(\omega) &= \int_0^\infty ds \exp(-i\omega s)\langle J_A(0) J_A(s)\rangle, \\ \Phi_3(\omega) &= \int_0^\infty ds \exp(-i\omega s)\langle J_{OH}(0) J_A(s)\rangle. \end{aligned} \tag{3.68}$$

There will exist two characteristic values of the ratio ξ_1/ξ_0, each of which will induce a value for the response ratio $\langle \dot{N}_A(t)\rangle/\langle \dot{N}_{OH}(t)\rangle$ numerically equal to that ξ_1/ξ_0 ratio itself. From Eqs. (3.67) one easily finds that these characteristic ratios are

$$\frac{\xi_1}{\xi_0} = R_\pm = \frac{1}{2\Phi_3}\{\Phi_2 - \Phi_1 \pm [(\Phi_2 - \Phi_1)^2 + 4(\Phi_3)^2]^{1/2}\}. \tag{3.69}$$

Of course R_+ and R_- in principle depend on ω, but they should approach limiting values as $\omega \to 0$, with leading-order corrections that are quadratic in ω.

Consider now the case of small ω, and suppose ξ_1/ξ_0 in $U(\mathbf{r}, t)$ is chosen to be one of the values $R_\pm(0)$. Then the system responds by slowly shifting its equilibrium to accommodate the slowly varying external field. The extent to which the system can follow the field depends on the relaxation time involved, and one such time belongs to each of R_+ and R_-. These relaxation times must be identified with the phenomenological times $\alpha_\mp^{-1}$, from Eq. (3.64). It is clear, furthermore, that the $R_\pm(0)$ should be identified with the coefficient $y(0)$ in Eq. (3.60).

To state the situation in other terms, the two linear combinations of concentration fluctuations

$$\left(\frac{\Delta N_{OH}}{V}\right) + R_\pm\left(\frac{\Delta N_A}{V}\right) \tag{3.70}$$

are just those that should exhibit single relaxation behavior. The corresponding combination of Eqs. (3.67),

$$\frac{d}{dt}\frac{\langle N_{OH}\rangle + R_\pm\langle N_A\rangle}{V} = -\frac{\beta\exp(i\omega t)\xi_0}{V}\{\Phi_1(\omega) + 2R_\pm\Phi_3(\omega) + R_\pm^2\Phi_2(\omega)\} \tag{3.71}$$

should therefore manifest a right member with ω dependence close to that for simple relaxation behavior ($K_\pm > 0$):

$$\frac{-i\omega\alpha_\pm}{K_\pm(\alpha_\pm + i\omega)}\left[\frac{\xi_0\exp(i\omega t)}{V}\right]. \tag{3.72}$$

In order to produce expressions for the relaxation times in terms of current correlation functions, it is necessary to expand both (3.71) and (3.72) through quadratic order in ω, and then compare corresponding terms. To the extent that separate relaxation times are indeed identifiable, ω dependence of $R_\pm$ should be negligible. Under that assumption we find the following molecular expressions for the chemical relaxation times:

$$\alpha_{\pm}^{-1} = \frac{\Phi_1^{(2)} + 2R_{\pm}(0)\Phi_3^{(2)} + R_{\pm}^2(0)\Phi_2^{(2)}}{2[\Phi_1^{(1)} + 2R_{\pm}(0)\Phi_3^{(1)} + R_{\pm}^2(0)\Phi_2^{(1)}]}; \tag{3.73}$$

we have introduced the correlation function moment quantities

$$\Phi_1^{(n)} = \int_0^{\infty} ds\, s^n \langle J_{\mathrm{OH}}(0) J_{\mathrm{OH}}(s) \rangle,$$

$$\Phi_2^{(n)} = \int_0^{\infty} ds\, s^n \langle J_{\mathrm{A}}(0) J_{\mathrm{A}}(s) \rangle,$$

$$\Phi_3^{(n)} = \int_0^{\infty} ds\, s^n \langle J_{\mathrm{OH}}(0) J_{\mathrm{A}}(s) \rangle. \tag{3.74}$$

The procedure outlined in Section III,D can readily be adapted to provide the quantum-mechanical version of expression (3.73).

If the molecular dynamics described by Eqs. (3.67) were precisely characteristic of a two-relaxation-process situation, then each of the Fourier transforms $\Phi_1(\omega)$, $\Phi_2(\omega)$, and $\Phi_3(\omega)$ would possess simple poles in the complex ω plane at $i\alpha_+$ and $i\alpha_-$. As we shall see in Section IV, however, this simple meromorphic character may not actually apply to the real aqueous solution case. Consequently, it is safer to rely on moment equations such as (3.73) to identify chemical relaxation rates.

It is important to keep in mind that rate coupling between the reactions (3.57) does not occur merely because both involve the same constituent H^+. The presence of HA and/or A^- in high concentration generally will affect the hydrogen-bond connectivity of the water network, and that in turn will affect the ability of the network to transport protons.

IV. Electrical Response

A. Frequency-Dependent Dielectric Function

Consider a region of empty space, for example that between parallel capacitor plates, which has an electrical field $\mathbf{E}_{\mathrm{ap}}$ varying harmonically with time:

$$\mathbf{E}_{\mathrm{ap}}(t) = \mathbf{E}_0 \exp(i\omega t). \tag{4.1}$$

If the region is then filled with homogeneous and isotropic matter (such as water or an aqueous solution), that matter then develops a polarization $\mathbf{P}$ which, in the linear response regime, may be related to $\mathbf{E}_{\mathrm{ap}}$ through the frequency-dependent dielectric function $\varepsilon(\omega)$:

$$4\pi\mathbf{P}(t) = \left[1 - \frac{1}{\varepsilon(\omega)}\right]\mathbf{E}_{ap}(t). \tag{4.2}$$

This polarization density causes the actual mean electrical field in the sample to be

$$\mathbf{E}(t) = \mathbf{E}_{ap}(t)/\varepsilon(\omega). \tag{4.3}$$

In general $\varepsilon(\omega)$ will be complex, and conventionally it is written in the following form:

$$\varepsilon(\omega) = \varepsilon'(\omega) - i\varepsilon''(\omega). \tag{4.4}$$

The imaginary part $-\varepsilon''$ represents dissipation. At very low frequency the dissipation in water and its solutions will be due to electrical conduction; this conductivity causes ε'' to diverge at $\omega = 0$:

$$\varepsilon''(\omega) = (4\pi\sigma/\omega) + 0(1), \tag{4.5}$$

where σ is the (low frequency) conductivity. As ω increases, absorbing regions of frequency are encountered which correspond in turn to molecular rotational relaxation ($\approx 10\ \text{cm}^{-1}$), to vibrational relaxation ($\approx 10^3\ \text{cm}^{-1}$), and finally to electronic excitation ($\approx 10^5\ \text{cm}^{-1}$).

The real and imaginary parts of $\varepsilon(\omega)$ are not independent, but are related to each other by the Kramers–Kronig equations (Kittel, 1958):

$$\varepsilon'(\omega) - \varepsilon'(\infty) = \frac{2}{\pi}\int_0^\infty \frac{u\varepsilon''(u)}{u^2 - \omega^2}\,du, \tag{4.6}$$

$$\varepsilon''(\omega) = -\frac{2\omega}{\pi}\int_0^\infty \frac{\varepsilon'(u) - \varepsilon'(\infty)}{u^2 - \omega^2}\,du. \tag{4.7}$$

Cauchy principal values for these integrals are to be taken at the integrand singularities. These Kramers–Kronig relations are simply a mathematical statement of causality, following from the requirement that medium response not precede the external perturbation.

The conductivity pole in ε'' at $\omega = 0$, shown in Eq. (4.5), necessarily affects ε' through Eq. (4.6). One easily demonstrates that near $\omega = 0$, ε' has to diverge to infinity as well. This merely reflects the ability of a conductor to shield completely any static electric field impressed upon it. However, pure water at least is a rather poor conductor ($0.04 \times 10^{-6}\ \Omega^{-1}\ \text{cm}^{-1}$ at 18°C), and so a convenient range of relatively low frequencies ($\approx 10^5$ Hz) is available over which a "static dielectric constant" ε_0 can experimentally be identified (as an attribute of water as a polar liquid consisting essentially of intact molecules) without significant interference from conductivity.

A satisfactory, completely general, theory of $\varepsilon(\omega)$ does not exist, which accounts for molecular motions, interactions, and polarizability. However some insight into the time-dependent dielectric behavior of water is possible through the theory restricted to rigid polar (but nonpolarizable) molecules. In this latter regime, $\varepsilon(\omega)$ may be calculated according to the following formula (Titulaer and Deutch, 1974):

$$\frac{[\varepsilon(\omega) - 1][2\varepsilon(\omega) + 1]\varepsilon_0}{(\varepsilon_0 - 1)(2\varepsilon_0 + 1)\varepsilon(\omega)} = -\int_0^\infty dt \exp(-i\omega t)\frac{d\phi(t)}{dt}. \tag{4.8}$$

In this expression, $\phi(t)$ stands for the normalized autocorrelation function of the moment $\mathbf{m}$ of a microscopically large, but macroscopically small, sphere of the substance, embedded in an infinite sea of its own kind.

$$\phi(t) = \langle \mathbf{m}(0) \cdot \mathbf{m}(t) \rangle / \langle m^2 \rangle. \tag{4.9}$$

The simplest reasonable form that one might expect for $\phi(t)$ is that for single exponential decay

$$\phi(t) = \exp(-t/\tau) \qquad (t \geq 0). \tag{4.10}$$

By inserting this form into Eq. (4.8) one concludes

$$\begin{aligned} \varepsilon(\omega) &= \frac{2\varepsilon_0^2 + i\omega\tau\varepsilon_0 - 1 + [Q(\varepsilon_0, \omega)]^{1/2}}{4\varepsilon_0(1 + i\omega\tau)}, \\ Q(\varepsilon_0, \omega) &= 4\varepsilon_0^4 + 4i\omega\tau\varepsilon_0^3 + (4 + 16i\omega\tau - 9\omega^2\tau^2)\varepsilon_0^2 \\ &\quad - 2i\omega\tau\varepsilon_0 + 1. \end{aligned} \tag{4.11}$$

Although this $\varepsilon(\omega)$ as a function of the complex variable ω is analytic in the neighborhood of the origin, it possesses characteristic singularities off the real axis. In particular, a simple pole resides at

$$\omega = i/\tau, \tag{4.12}$$

and a pair of square-root branch points exist at the roots of $Q(\varepsilon_0, \omega)$:

$$\omega = \frac{i}{9\varepsilon_0\tau}\{2\varepsilon_0^2 + 8\varepsilon_0 - 1 \pm 2[\varepsilon_0^4 + 8\varepsilon_0^3 + 6\varepsilon_0^2 - 13\varepsilon_0 - 2]^{1/2}\}. \tag{4.13}$$

Thus we see that a simple $\phi(t)$ can give rise to a relatively complicated dielectric function.

Expression (4.11) for $\varepsilon(\omega)$ should be compared with the elementary "single relaxation" form that is often quoted (Fröhlich, 1949):

$$\varepsilon(\omega) = \frac{\varepsilon_0 + i\omega\tau}{1 + i\omega\tau}, \tag{4.14}$$

which exhibits only the simple pole at i/τ. Obviously Eq. (4.14) requires a more elaborate $\phi(t)$ than the one shown in Eq. (4.10).

Considerable theoretical interest has recently been devoted to "hydrodynamic tails" in correlation functions that determine transport and kinetic properties (Alder and Wainwright, 1970). In the case of angular momentum for a single molecule, local vorticity created by that molecule in its surroundings causes the autocorrelation function to display a $t^{-5/2}$ tail at long time t (Ailawadi and Berne, 1971). Likewise, the rank-one orientational correlation functions for a single molecule decay to zero at long time as $t^{-7/2}$ (Garisto and Kapral, 1974). We take these facts as evidence that $\phi(t)$ in Eq. (4.9) also should display a long-time inverse-power tail since, after all, the local moment $\mathbf{m}(t)$ is additively composed of separate molecular contributions. In particular, we would expect $t^{-7/2}$ to be the relevant behavior for $\phi(t)$. Then the Fourier transform quantity on the right side of the $\varepsilon(\omega)$ formula (4.8) would possess branch-point character at the origin of the complex ω plane, associated with occurrence of $\omega^{7/2}$ in a small-ω expansion. Evidently $\varepsilon(\omega)$ itself will have a branch point at $\omega = 0$.

Obviously liquid water as a dielectric substance is much more complicated than the assumptions behind Eq. (4.8) would permit. Nevertheless, the same features apply in water that cause one to conclude from Eq. (4.8) that $\varepsilon(\omega)$ contains a branch-point singularity at the origin. The resulting dielectric behavior is obviously relevant to the motion of ions in water, particularly H^+ and OH^-.

B. Electrical Conduction

The time-correlation function expression for the electrical conductivity σ has the following form (Zwanzig, 1965):

$$\sigma = \beta \int_0^\infty dt \langle J_x(0) J_x(t) \rangle. \tag{4.15}$$

J_x is the x-component of the total electrical current flowing in the system. This formula refers specifically to a system with unit volume subject to periodic boundary conditions. In the case of pure water, J_x obviously comprises contributions both from H^+ and OH^-; aqueous solutions may contain other charge carriers as well.

We have noted earlier that the high static dielectric constant ε_0 aids in production of carriers in water, owing to the negative solvation free energy of H^+ and OH^-. At the same time the frequency dependence in $\varepsilon(\omega)$ causes a retardation of the motion of these ions, since local polarization has difficulty in following ion drift. Temporal persistence of spontaneous current fluctuations at equilibrium is therefore damped somewhat, and the correlation function expression (4.15) reflects that fact by yielding a correspondingly reduced σ value.

One special aspect of the dynamics of local polarization as it follows H^+ and OH^- motion is clear from Figs. 1 and 2. The sequential proton exchanges along hydrogen-bond chains (which produce high apparent mobilities) automatically switch the orientation of water molecules in a favorable way, without the necessity for real molecular rotation. The polarization lag that retards forward motion will therefore largely reside in regions to the side of the conductive chain.

In the case of a spherical ion of radius b moving through a fluid dielectric continuum, Zwanzig has calculated the dielectric friction (Zwanzig, 1970). His calculation assumed that the dielectric dispersion could be represented by a single relaxation time τ_d. The result for the dielectric friction constant ζ_d was the following:

$$\zeta_d = \frac{Kq^2(\varepsilon_0 - \varepsilon_\infty)\tau_d}{b^3\varepsilon_0(2\varepsilon_0 + 1)}. \tag{4.16}$$

Here q stands for the charge on the moving ion, and the constant K depends on the hydrodynamic boundary conditions applied at the surface of the ion:

$$\begin{aligned} K &= 3/8 \qquad \text{(sticking)} \\ &= 3/4 \qquad \text{(slipping).} \end{aligned} \tag{4.17}$$

Although it may seem paradoxical that slipping conditions produce greater dielectric friction than sticking conditions, it should be realized that the former permits nearby liquid to flow faster around the moving ion, thereby creating a greater polarization lag.

Obviously, there are grave difficulties in attempting to apply Zwanzig's result (4.16) to H^+ or OH^- in water, because of the exchange character of motion for these ions. Furthermore, it is unclear what an appropriate choice for the radius b would be for these species. However the fact that ζ_d is essentially proportional to τ_d/ε_0 is suggestive, and perhaps not far from the truth for H^+ and OH^- in water. This ratio declines monotonically between 0° and 75°C by a factor of about 3.9. No doubt a substantial part of the rising mobilities (equivalent conductances) with increasing temperature of H^+ and OH^- shown in Table II can be attributed to this declining dielectric friction.

Another important ingredient in understanding temperature variations of H^+ and OH^- mobilities is the topological nature of hydrogen-bond connections between molecules, and the rate at which such connections break and reform in new patterns. Eigen's description of proton conduction and diffusion in water (Eigen, 1964) seems to demand the interpretation that the liquid consists of disconnected finite (i.e., independent of system size) clusters of bonded water molecules, with stochastic repositioning of nearby water molecules to attach to the edge of a cluster acting as rate-determining events. However, recent molecular dynamics simulations (Geiger *et al.*, 1977) show instead that an extended space-filling network picture for liquid water is clearly preferable, when an energy definition of "hydrogen bond" is chosen which conforms to a wide range of other physical and chemical measurements and properties. To reconcile the apparent contradition, Eigen's hydrogen bonds would have to be restricted to greater binding energy and/or shorter O–O separation than required otherwise.

Figure 12 shows that even if the hydrogen bonds involved are strong

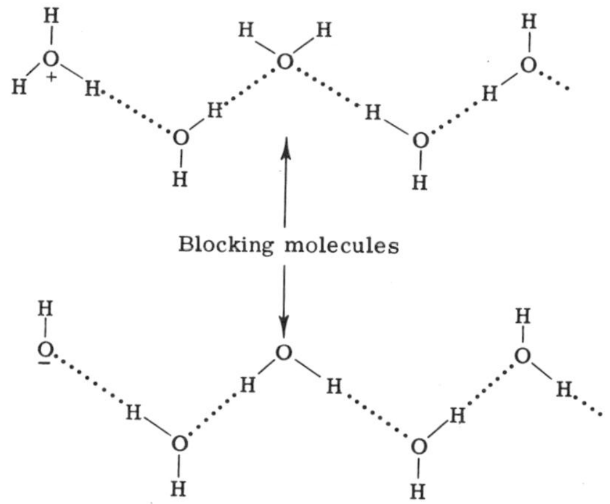

Fig. 12. Interrupted exchange pathways. Sequential proton jumps along a hydrogen-bond path for H^+ and OH^- transport can be blocked by a water molecule acting, respectively, as double proton acceptor (top) or double proton donor (bottom).

enough, a chain of bonds is not sufficient to guarantee the possibility of transport. All of these bonds must in fact be properly oriented beforehand. H^+ motion along a chain can effectively be blocked if a water molecule acting as a double proton acceptor is incorporated in that chain (Fig. 12, top). Likewise, a water molecule which acts as a double donor in a chain can effectively block passage of OH^- (Fig. 12, bottom). Evidently both formation and destruction of strong hydrogen bonds, as well as reorientations to shift "blocking molecules" are important in electrical conduction and ionic diffusion of H^+ and OH^- in water.

It is inherent in the crystallographic structure of ice that each molecule has many hydrogen bond paths passing through it. Considering any molecule and pairs of its four bonded neighbors, one sees that the chosen molecule simultaneously serves once as double proton donor, once as double proton acceptor, and four times as a nonblocking molecule (Hankins *et al.*, 1970). The multiplicity of hydrogen-bond pathways in the crystal thus can permit blockages to be bypassed. Set against this topological advantage is the rigidity of the crystal that would inhibit formation of those very short and strong hydrogen bonds necessary for charge transport. Evidently, this latter feature dominates, for mobilities of H^+ and OH^- are less in ice than in liquid water, as noted earlier (Section I,B).

Owing to the relative magnitudes of three-molecule nonadditivity in water-molecule interactions, incorporation of "double donor" and "double acceptor" molecules in hydrogen bond chains is energetically unfavorable (Hankins *et al.*, 1970). Of these two, the double donor is the configuration more costly in energy. Unlike ice, the incompletely bonded liquid has a choice in how many blocking sites exist; therefore relatively fewer double donor occurrences than double acceptor occurrences are expected (the non-additivity energy difference between them is approximately 0.5 kcal/mole.) Thus, blockage should more frequently occur for H^+ motion than for OH^- motion. Obviously, then, the fact exhibited in Table II that H^+ is more mobile than OH^- must stem from yet another feature; and indeed this feature must be sufficiently potent more than just to overcome the occurrence probabilities for blocking molecules.

The limiting equivalent conductances (Table II) imply the following activation energies for electrical charge transport at 20°C:

$$\begin{aligned} &2.59 \text{ kcal/mole } (H^+), \\ &3.11 \text{ kcal/mole } (OH^-). \end{aligned} \tag{4.18}$$

It was stressed earlier in Sections II,B and C that hydrogen bonds in the immediate vicinity of a solvated H^+ tend to be stronger and shorter than those for a solvated OH^-. The difference in activation energies (4.18) indicates greater difficulty for OH^- compared to H^+ in reaching a "transition state" between successive "natural" configurations. One contributing factor to the difference is the presence of proton double minima for OH^- hydrogen bonds, in contrast to flat proton potentials for corresponding solvated H^+ structures. Another, and probably more important, factor is that solvated H^+ is surely surrounded on the average by a larger dendritic array of favorable hydrogen bonds than is solvated OH^-, owing to the tendency toward stronger bonds with the former. Consequently the H^+ hydration complex finds at its perimeter a greater set of configurational opportunities for easy proton transfer in the field direction and among this greater set will occur more opportunities for low-barrier transfers.

Notice that the ratio of mobilities of H^+ and OH^- tends toward unity as the temperature rises. Presumably this is associated with thermal disruption of the respective solvation complexes. It would be useful to know by direct measurement if pressure rise has a similar effect.

We have noted that ionic motion in water is affected by coupling to dielectric dispersion, and that the frequency-dependent dielectric function $\varepsilon(\omega)$ exhibits branch-point characteristics in the complex ω plane. All chemical reactions in water involving proton transfer are therefore inevitably coupled to dielectric response and therefore partake of complex-ω branch-point behavior. This is the reason behind earlier cautioning (Section III,E) that the chemical current correlation transforms $\Phi_1(\omega)$, $\Phi_2(\omega)$, and $\Phi_3(\omega)$ in Eq. (3.68) should not cavalierly be treated as meromorphic functions.

C. Wien Effect

Wien was the first to point out that electrolytes display a characteristic non-Ohmic behavior; namely, their electrical conductance increases with increasing electrical field (Wien, 1928, 1931). This phenomenon exists for both strong and weak electrolytes. In the case of the former, the strong electrical field destroys the normal ion atmosphere, which exerts a retarding effect on ion mobility at small fields. In the case of the latter, strong fields enhance the rate of ionization without producing a compensating change in association rate. As a consequence, the weak electrolyte's ionization constant increases.

In comparison, the relative change in conductance at high fields is much greater for weak electrolytes than for strong electrolytes.

Onsager has worked out the theory of the Wien effect for weak electrolytes (Onsager, 1934). He concludes that the dissociation constant K_d should exhibit the following dependence on electric field strength E:

$$\frac{K_d(E)}{K_b(0)} = \frac{J_1[4(-\zeta_0 q)^{1/2}]}{2(-\zeta_0 q)^{1/2}}$$

$$= 1 + 2\zeta_0 q + \frac{(4\zeta_0 q)^2}{2!3!} + \frac{(4\zeta_0 q)^3}{3!4!} + \cdots, \tag{4.19}$$

where for a 1–1 electrolyte

$$q = \frac{e^2}{2\varepsilon_0 k_B T}, \qquad \zeta_0 = \frac{|eE|}{2k_B T}. \tag{4.20}$$

J_1 is the usual Bessel function which, for the imaginary argument shown, is monotonic. It is noteworthy that Onsager's result (4.19) depends on the absolute value of the field in such a way that the leading order effect is proportional to $|E|$ itself. In fact, that linear field effect would actually not apply at very small fields, owing to neglect of terms in derivation of Eq. (4.19). However, in the case of water, these neglected terms (associated with ion atmosphere perturbation) are appreciable only at very small fields indeed.

With an electrical field of 100 kV/cm in water at 25°C, one has

$$q = 3.5790 \times 10^{-8} \text{ cm},$$

$$\zeta_0 = 1.9462 \times 10^{6} \text{ cm}^{-1},$$

$$K_d(E)/K_d(0) = 1.146, \tag{4.21}$$

which incorporates only slight deviation from Onsager's linear field effect regime.

Eigen and DeMaeyer have employed the Wien effect in pure water to study the rate of the fundamental recombination reaction (1.1) (Eigen and DeMaeyer, 1955). They used field strengths in the range 50–100 kV/cm, with rise times less than 10^{-8} sec to generate extra H^+ and OH^- ions in the liquid. The relaxation engendered by the subsequent neutralization reaction could then be monitored as electrical conductivity change.

V. Conclusions

The present exposition provides a formal theoretical framework for proton transfer reactions and kinetics in water. A major advantage of such a formalism is that it helps point out areas of incomplete information. Consequently, it seems highly desirable to pursue the following in the near future.

1. More detailed and accurate quantum-mechanical studies of potential energy surfaces for solvated H^+ and OH^- ions.
2. Monte Carlo and molecular dynamics investigation of H^+ and OH^- ions in water.
3. Experimental study of dielectric properties and H^+ and OH^- mobilities in supercooled water, and in water at elevated pressure.
4. Development of experimental techniques to study dissociation relaxation down to the picosecond range.
5. Measurement of dissociation relaxation in H_2O–D_2O mixtures.
6. Investigation of models for proton (and proton hole) transfer in random networks.

These projects, if successfully carried to completion, would substantially improve quantitative understanding of the subject of this review.

References

Ailawadi, N. K., and Berne, B. J. (1971). *J. Chem. Phys.* **54**, 3569–3571.
Alder, B. J., and Wainwright, T. E. (1970). *Phys. Rev. A.* **1**, 18–21.
Almlöf, J. (1972). *Acta Crystallogr., Sect. B* **28**, 481–485.
Almlöf, J., and Wahlgren, U. (1973). *Theor. Chim. Acta* **28**, 161–168.
Bell, R. A., Christoph, G. G., Fronczek, F. R., and Marsh, R. E. (1975). *Science* **190**, 151–152.
Benedict, W. S., Gailar, N., and Plyler, E. K. (1956). *J. Chem. Phys.* **24**, 1139–1165.
Bernal, J. D., and Fowler, R. H. (1933). *J. Chem. Phys.* **1**, 515–548.
Berry, R. S. (1969). *Chem. Rev.* **69**, 533–542.
Beurskens, G., and Jeffrey, G. A. (1964). *J. Chem. Phys.* **41**, 924–929.
Chamberlain, J. W., and Roesler, F. L. (1955). *Astrophys. J.* **121**, 541–547.
Chambers, R. D. (1973). "Fluorine in Organic Chemistry," p. 65. Wiley, New York.
Clough, S. A. (1976). Personal communication.
Collie, C. H., Hasted, J. B., and Ritson, D. M. (1948). *Proc. Phys. Soc. (London)* **60**, 145–160.
Covington, A. K., Robinson, R. A., and Bates, R. G. (1966). *J. Phys. Chem.* **70**, 3820–3824.
Darwent, B. deB. (1970). *Natl. Stand. Ref. Data Ser., Nat. Bur. Stand.* **31**, p. 41.
Debye, P. (1942). *Trans. Electrochem. Soc.* **82**, 265–271.
DePas, M., Leventhal, J. J., and Friedman, L. (1968). *J. Chem. Phys.* **49**, 5543–5544.
Dogonadze, R. R., Kharkats, Yu. I., and Ulstrup, J. (1976). *Chem. Phys. Lett.* **37**, 360–364.
Dyke, T. R., and Muenter, J. S. (1973). *J. Chem. Phys.* **59**, 3125–3127.
Eigen, M. (1964). *Angew. Chem.* **3**, 1–18.
Eigen, M., and DeMaeyer, L. (1955). *Z. Elektrochem.* **59**, 986–993.
Eigen, M., and DeMaeyer, L. (1958). *Proc. R. Soc. (London) Ser. A* **247**, 505–533.
Eisenberg, D., and Kauzmann, W. (1969). "The Structure and Properties of Water." Oxford Univ. Press, New York.
Fieser, L. F., and Fieser, M. (1956). "Organic Chemistry," p. 163. Reinhold, New York.
Finney, J. L. (1977). *Proc. R. Soc. (London) Ser. B* **278**, 3–31.
Fletcher, N. H. (1970). "The Chemical Physics of Ice." Cambridge Univ. Press, London and New York.
Fröhlich, H. (1949). "Theory of Dielectrics," p. 72. Oxford Univ. Press, London.
Garisto, F., and Kapral, R. (1974). *Phys. Rev. A.* **10**, 309–318.
Geiger, A., Rahman, A., and Stillinger, F. H. (1978). *J. Chem. Phys.*, to be published.
Grunwald, E., Loewenstein, A., and Meiboom, S. (1957). *J. Chem. Phys.* **27**, 630–640.
Hamann, S. D. (1963). *J. Phys. Chem.* **67**, 2233–2235.
Hankins, D., Moskowitz, J. W., and Stillinger, F. H. (1970). *J. Chem. Phys.* **53**, 4544–4554; Erratum (1973), **59**, 995.
Hepler, L. G., and Woolley, E. M. (1973). *In* "Water, A Comprehensive Treatise" (F. Franks, ed.), Vol. 3, p. 153. Plenum, New York.
Herzberg, G. (1950). "Molecular Spectra and Molecular Structure. I. Spectra of Diatomic Molecules," p. 560. Van Nostrand, Princeton, New Jersey.
Hill, T. L. (1956). "Statistical Mechanics." McGraw-Hill, New York.
Janoschek, R., Preuss, H., and Diercksen, G. (1967). *Int. J. Quantum Chem.* **1**, 649–652.
Jeffrey, G. A., Jordon, T. H., and McMullan, R. K. (1967). *Science* **155**, 689–691.
Jones, W. M. (1952). *J. Am. Chem. Soc.* **74**, 6065–6066.
Jones, W. M. (1968). *J. Chem. Phys.* **48**, 207–214.
Kebarle, P., Searles, S. K., Zolla, A., Scarborough, J., and Arshadi, M. (1967). *J. Am. Chem. Soc.* **89**, 6393–6399.
Kell, G. S., McLaurin, G. E., and Whalley, E. (1968). *J. Chem. Phys.* **49**, 2839–2842.

Kirkwood, J. G. (1939). *J. Chem. Phys.* **7**, 911–919.
Kittel, C. (1958). "Elementary Statistical Physics," pp. 206–210. Wiley, New York.
Kollman, P. A., and Bender, C. F. (1973). *Chem. Phys. Lett.* **21**, 271–274.
Kotowski, A., ed. (1964). "Gmelins Handbuch der Anorganischen Chemie. Sauerstoff," Lieferung 6, System-Nummer 3, p. 1994. Verlag Chemie, Heidelberg.
Kraemer, W. P., and Diercksen, G. H. F. (1970). *Chem. Phys. Lett.* **5**, 463–465.
Kutz, H. D., Oppenheim, I., and Ben-Reuven, A. (1974). *J. Chem. Phys.* **61**, 3313–3326.
Lie, G. C., and Clementi, E. (1975). *J. Chem. Phys.* **62**, 2195–2199.
Long, F. A., and Purchase, M. (1950). *J. Am. Chem. Soc.* **72**, 3267–3273.
Long, F. A., Dunkle, F. B., and McDevit, W. F. (1951). *J. Phys. Colloid Chem.* **55**, 829–842.
Lundgren, J.-O. and Olovsson, I. (1967). *Acta Crystallogr.* **23**, 966–971, 971–976.
Lundgren, J.-O., and Olovsson, I. (1968). *J. Chem. Phys.* **49**, 1068–1074.
McMullan, R. K., Mak, T. C. W., and Jeffrey, G. A. (1966). *J. Chem. Phys.* **44**, 2338–2345.
Malmberg, C. G., and Maryott, A. A. (1956). *J. Res. Natl. Bur. Stand.* **56**, 1–8.
Mills, R. (1973). *J. Phys. Chem.* **77**, 685–688.
Narten, A. H. (1972). *J. Chem. Phys.* **56**, 5681–5687.
Narten, A. H., and Levy, H. A. (1971). *J. Chem. Phys.* **55**, 2263–2269.
Newton, M. D., and Ehrenson, S. (1971). *J. Am. Chem. Soc.* **93**, 4971–4990.
Olovsson, I. (1968). *J. Chem. Phys.* **49**, 1063–1067.
Onsager, L. (1934). *J. Chem. Phys.* **2**, 599–615.
Onsager, L. (1973). *In* "Physics and Chemistry of Ice" (E. Whalley, S. J. Jones, and L. W. Gold, eds.), p. 10. Royal Soc. of Canada, Ottawa.
Peterson, C., and Pfeiffer, G. V. (1972). *Theor. Chim. Acta* **26**, 321–330.
Popkie, H., Kistenmacher, H., and Clementi, E. (1973). *J. Chem. Phys.* **59**, 1325–1336.
Pople, J. A., Schneider, W. G., and Berstein, H. J. (1959). "High-Resolution Nuclear Magnetic Resonance," Chapter 18. McGraw-Hill, New York.
Rahman, A., and Stillinger, F. H. (1973). *J. Am. Chem. Soc.* **95**, 7943–7948.
Résibois, P. M. V. (1968). "Electrolyte Theory," p. 33. Harper, New York.
Robinson, R. A., and Stokes, R. H. (1959). "Electrolyte Solutions," p. 544, Butterworths, London.
Smith, D. F., and Overend, J. (1972). *Spectrochim. Acta* **28A**, 471–483.
Stillinger, F. H. (1975). *Adv. Chem. Phys.* **31**, 1–101.
Stillinger, F. H. (1977). *Proc. R. Soc. (London) Ser. B* **278**, 97–110.
Stillinger, F. H., and Lovett, R. (1968). *J. Chem. Phys.* **49**, 1991–1994.
Stillinger, F. H., and Rahman, A. (1974). *J. Chem. Phys.* **60**, 1545–1557.
Stoeckenius, W. (1976). *Sci. Am.* **234**, No. 6, 38–46.
Stokes, R. H., and Mills, R. (1965). *In* "International Encyclopedia of Physical Chemistry and Chemical Physics. Topic 16: Transport Properties of Electrolytes" (E. A. Guggenheim, J. E. Mayer, and F. C. Tompkins, eds.), Vol. 3, p. 74. Pergamon, New York.
Swalen, J. D., and Ibers, J. A. (1962). *J. Chem. Phys.* **36**, 1914–1918.
Titulaer, U. M., and Deutch, J. M. (1974). *J. Chem. Phys.* **60**, 1502–1513.
Tödheide, K. (1972). *In* "Water, A Comprehensive Treatise" (F. Franks, ed.), Vol. 1, pp. 463–513. Plenum, New York.
Van Raalte, D., and Harrison, A. G. (1963). *Can. J. Chem.* **41**, 3118–3126.
Vereschagin, L. F., Yakovlev, E. N., and Timofeev, Yu. A. (1975). *JETP Lett.* **21**, 304–305.
Vidulich, G. A., Evans, D. F., and Kay, R. L. (1967). *J. Phys. Chem.* **71**, 656–662.
Weast, R. C., ed. (1975). "Handbook of Chemistry and Physics," 56th edition, pp. D136–D139. Chemical Rubber Co., Cleveland, Ohio.
Wiberg, K. B. (1964). "Physical Organic Chemistry," p. 252. Wiley, New York.
Wien, M. (1928). *Phys. Z.* **29**, 751–755.

Wien, M. (1931). *Phys. Z.* **32**, 545–547.
Young, T. F., Maranville, L. F., and Smith, H. M. (1959). *In* "The Structure of Electrolytic Solutions" (W. J. Hamer, ed.), pp. 35–63. Wiley, New York.
Zwanzig, R. W. (1965). *Annu. Rev. Phys. Chem.* **16**, 67–102.
Zwanzig, R. W. (1970). *J. Chem. Phys.* **52**, 3625–3628.

Subject Index

D

E

F

G

H

I

A
B
C 8
D 9
E 0
F 1
G 2
H 3
I 4
J 5